OPTIMIZING RESERVOIR RESOURCES

OPTIMIZING RESERVOIR RESOURCES
Including a New Model for Reservoir Reliability

Charles ReVelle

JOHN WILEY & SONS, INC.

New York / Chichester / Weinheim / Brisbane / Singapore / Toronto

This book is printed on acid-free paper. ♾

Published simultaneously in Canada.

This publication is designed to provide accurate and authoritative information in regard to the subject matter covered. It is sold with the understanding that the publisher is not engaged in rendering professional services. If professional advice or other expert assistance is required, the services of a competent professional person should be sought.

Library of Congress Cataloging-in-Publication Data:

ReVelle, Charles.
   Optimizing reservoir resources : including a new model for reservoir reliability / Charles ReVelle.
     p. cm.
   Includes bibliographical references and index.
   ISBN 0-471-18877-8 (cloth : alk. paper)
   1. Reservoirs—Mathematical models. 2. Water-supply. 3. Flood control. 4. Water-power. I. Title.
TD395.R48 1999
333.91′15—dc21   98-38927

Printed in the United States of America.

10 9 8 7 6 5 4 3 2 1

To Penny,
  Water and love are twins
  Without both, life withers

CONTENTS

ACKNOWLEDGMENTS

Books come about for numerous reasons; each has its own story of motivation. This book is the result of an activity I undertook in 1993 as a Visiting Scholar at the Institute of Water Resources (IWR) of the U.S. Army Corps of Engineers. I was invited by Eugene Stakhiv, who heads the Policy and Special Studies Division of IWR. His charge for me was to rationalize the reallocation process for reservoirs. I soon realized that the process was made especially difficult by the presence of a number of problems that remained unsolved in the general sense, and that in order to make progress on reallocation, I first needed to make progress on reexamining some fundamental water resource problems. Gene realized this and let me proceed with the task as I perceived it. I am enormously grateful to him for encouraging me to tackle these problems and for giving me the freedom to concentrate my efforts on them.

If this book proves to be an enduring contribution to water resources management, it will in no small measure be due to the support and encouragement that I received at the Institute for Water Resources. I had numerous thoughtful and congenial colleagues with whom I was able to work and who were willing to share their expert knowledge on the conceptual and practical issues that confront water resource decision makers. I would particularly like to note the intellectual contributions of Bill Werick, Ted Hillyer, Bob Brumbaugh, and Kyle Schilling, the Director of IWR. It was a pleasure to participate with them in an exploration of ideas on the problems of water resources management confronting society today. IWR is the Institution that regularly transforms concepts to practice, and it is a tribute to the leadership of such individuals as Gene Stakhiv and Kyle Schilling that IWR has become a leader in implementing the best innovations in water resources and environmental management.

INTRODUCTION

The focus of this book is the design and operation of the surface water reservoir and systems of reservoirs. Water resources is an area of study rich in problems and challenges. Further, the mathematical form used to model the reservoir shows up again and again in numerous industrial and commercial settings where the management of inventory is at issue. Thus, study of the reservoir model reveals not only the subject itself, but illustrates a basic mathematical structure that can be utilized in many other applied settings.

The fundamental components that make up surface water systems include not only reservoirs, but their withdrawal structures and spillways as well as associated pipelines, irrigation channels, and hydropower units, components that reflect the multiple functions of the reservoirs. These functions include water supply for municipal and industrial use, irrigation, flood control, recreation, hydropower, and flow maintenance for navigation or aquatic life. All these functions will be modeled in this book, although water supply, flood control, and hydropower are emphasized.

The water reservoir takes different structural forms depending on its design functions. If the reservoir is used only for flood control, it can be almost as simple in design as a bathtub, with a single, limited-size, unadjustable outlet structure as well as a spillway. Alternatively, the reservoir can deliver water through multiple hydropower units or through complex units such as a submerged tower with inlet ports at multiple levels of the reservoir. Such a tower might be used to mix water for temperature control or even oxygen control.

The reservoir, for our purposes, will be an impoundment created by either an earthen dam or a concrete dam. It will have a spillway to get rid of excessively high and unexpected flows—if it should become absolutely necessary to do so. And it will have one or more outlet structures, at least one of which discharges water to the stream below the reservoir, possible through gates, possibly through a turbine-generator.

The water discharged to the stream may be a planned release to the stream to maintain desired aquatic life or navigational flows, or the discharge may be used later for water supply withdrawal. Alternatively, this water released to the stream may be "wasted" from the reservoir through its gates to keep the reservoir sufficiently empty to prevent use of the spillway. Since most spillways are never tested (for fear of their destruction), this latter practice is only common sense. If the water used for water supply or irrigation is not drawn from the stream but from the reservoir, a separate structure will provide the means of withdrawal. The structure might be a submerged port connected to a pipeline that directs water to municipal and industrial use.

This basic reservoir structure provides the flexibility needed to begin to model the arrival of water at the reservoir, its storage in the reservoir, and its dispatch to various uses. Alternatively, this template of uses can be used to model the design process that determines a required reservoir capacity.

The forms that we focus on in this book fall in two categories: deterministic and stochastic models. The simplest of the deterministic models are the linear programming models that choose optimal releases or capacities and simulation models that experiment with various releases and capacities. Both model types deal with known inflows to the reservoir. Interestingly, the stochastic form we investigate in Chapter 7 reduces to the linear programming model as well.

Indeed, we are fortunate that many of the hydrologic optimization models that have been built fall naturally in the category of linear programming or into categories strongly related to linear programming. We are fortunate because these optimization models are the simplest to explain and the simplest to solve. We unapologetically emphasize linear programming models in this book because of the ease of problem statement and ease of solution. The analyst does not need to write a computer code for the solution of any particular linear programming model; multiple, easy-to-use solvers adaptable to all linear programming problems are already available in the marketplace. Further, solutions obtained are optimal to the problems at hand. Alternatives derived under multiple objectives can be generated easily. Finally, a linear programming problem statement is as much an explanatory device as it is a means of solution.

The plan of this book is to divide reservoir models into two parts. In Part I, we begin with relatively established technology (models) for the design or operation of the surface water reservoir. The presentation will be by function, beginning with water supply, proceeding to flood control, and then hydropower. These three functions will be integrated in a model of a single reservoir operating in a deterministic environment. Finally, in this opening part of the book, irrigation as well as in-stream releases will be added to the list of functions. Through most of the discussion in Part I, it will be assumed that a division or allocation of reservoir services between these functions has already taken place.

In Part II, which is Chapters 6, 7, 8, and 9, we relax the assumption of prior division of services between the functions and show how, in both a deterministic and random environment, to allocate services between functions for new and existing reservoirs that provide water supply, flood control, and hydropower. It is in Chapter 6 that we encounter the critical need for a new model of a water supply reservoir, a model that can in an efficient fashion ensure the reliability of the water supply function. Such a model is created and demonstrated in Chapter 7. In Chapters 8 and 9, the new model is folded into the allocation and reallocation of reservoir services.

PART I

DETERMINISTIC MODELS FOR THE OPERATION AND DESIGN OF RESERVOIRS

Part I of the book, consisting of Chapters 1 through 5, considers deterministic models for the operation and design of reservoirs and systems of reservoirs. The models that are discussed in Part I are generally derivative of models that are already in the literature, but in some instances do introduce new methodology. Chapter 1 examines the basic and overarching linear programming model for the water supply reservoir. This model forms the base for nearly all the resevoir models discussed in the remainder of the book. The mechanism of evaporation is added to the model's structure as a first modification to its basic format. Then the formulation structure to consider flood control is introduced. Chapter 1 closes with a simulation model and the fundamental reservoir operating rules that are used to guide the steps of the simulation.

The water supply function is extended to multiple reservoirs in Chapter 2. The challenge in this chapter is to determine the extent to which each reservoir should contribute to the overall needs of the system. The issue is examined in the context of both an existing system of reservoirs and a system that is yet to be built. Studying the existing system first, we can see that the issue is complicated by the question of whether the contribution of each reservoir should be fixed for a particular month of the year or should be allowed to vary, based on the system storages or flow conditions of the moment. Pure linear programming models to calculate the contributions are offered once again, but, in addition, it is shown that optimization models can be embedded within simulation models and vice versa. Methods to develop real-time nonformulaic decisions, which are adapted to and special to the conditions at hand at a particular moment in time, are offered as well. The last portion of the chapter takes up the more difficult question of how simultaneously to design the system (determine storage capacities) and determine operating rules for the individual reservoirs.

The hydropower function is taken up in Chapter 3 where we show how to find the smallest reservoir capable of delivering a specified level of hydroelectric energy. That is, we show how to determine a storage-yield curve where in this case the yield is energy. Prices for electrical energy are then introduced to the equation and revenues rather than firm energy yield are maximized. Finally, the hydropower function is extended to multiple reservoirs.

Chapter 4 takes up two engaging issues: how to allocate the costs to joint participants in water supply projects and how to sequence the construction of projects that will not be built all at once. The cost allocation issue is investigated both in simple water supply only situations with multiple participants as well as in situations where a reservoir serves multiple functions. Time staging or sequencing is seen as a problem in cost minimization, and both a linear integer program as well as an exact special-case algorithm are offered. The difficult confluence of these two problems is discussed.

All of the methods for operation and design are brought back in reprise in Chapter 5 where previous functions are recognized as constraints on operation and new functions are introduced. These new functions are seen as goals for satisfactory operation. They include irrigation, release to the stream, and reservoir storage for recreation. They are treated as goals in the sense that the functions are to be performed as well as possible, perhaps with targets that it would be beneficial to hit or to be near. Hard constraints, however, constraints that cannot be violated, are absent for these several functions. Interesting mathematical "tricks" are introduced for optimizing over these functions.

CHAPTER 1

THE WATER SUPPLY FUNCTION: SINGLE RESERVOIR

In order to create an initial model for reservoir operation or design, we need a record of streamflow inputs to the reservoir. The record of inflows that is selected for utilization reflects the intended functions that are to be considered in the optimization model. For instance, if the model will size a reservoir to be used only for a water supply, the worst records of monthly inflows, usually no more than three or four years' worth, must be included within the record used for analysis. The worst record of inflows is referred to as the "critical period."

Water demand and water supplies (inflows) are not matched in time—peak demands often correspond to seasons when inflows (and rainfall) are least. This is the reason that water supply dams are built—to store water from seasons of abundance for seasons of shortage. And water demands and water supplies are not matched spatially. This is the reason that projects with pipelines and diversions are built—to provide water from areas that are water-rich (if only temporarily) to areas where water is in short supply (if only temporarily).

WATER SUPPLY WITH CONSTANT RELEASE

The initial model presumes a steady release to the water supply month after month. A second model considers a consistent year-to-year release for each January, a consistent release for each February, and so on. We introduce the following notation to structure the basic reservoir model.

t, n = index of months and total length in months of the critical period

s_t = storage at the end of month t, billion gallons, unknown

s_o = starting end-of-period storage, unknown

q = steady month-to-month release for water supply, billion gallons, specified in this first model, later a variable

w_t = water wasted to the stream from the reservoir in month t for lack of storage space, referred to as spill, billion gallons, unknown

c = capacity of the reservoir to store water, billion gallons, unknown in this first model, later a variable

i_t = historically recorded streamflow into the reservoir, month t, of the critical period, billion gallons, known

The initial model seeks the smallest reservoir capacity needed to sustain a steady release of q throughout the duration of the critical period, at the same time requiring the ending reservoir condition to be no worse than the condition at the start of the critical period.

What is the critical period? That is, what is the beginning month and year and ending month and year of the critical record? Speaking generally, it is a several-year sequence of flows far lower than normal. To be more specific requires specification of the value of the steady water supply release q. With this number given, the critical period is the set of months of least length for which expansion to include earlier or later months in the sequence produces no change in the needed reservoir capacity. The critical period is not fixed and it is not the same set of months for all levels of water supply. In general, as the steady water supply q is increased, the critical period expands to encompass more of the historical record. Each unit of increase in the value of q does not result in expansion of the critical period, although it will require added reservoir capacity. Rather, expansion occurs at certain discrete values. Between successive pairs of discrete values, as q increases, the needed reservoir capacity increases linearly.

If the critical period is not fixed in length, how can the water resources engineer settle on the months to use without undertaking another layer of analysis? The answer is to use a low flow record of sufficient length that the critical period is sure to be contained within it. Extension of that record in any direction will not change the required capacity. The record may be longer than is absolutely necessary, but no further analysis of record length is needed. Once the linear programming model is structured, it can be seen that the entire length of the record could be used without stretching the limits of most commercial linear programming codes and modern desktop computers.

The linear programming problem statement, originally due to Dorfman (1965) in nonstandard form is

Minimize $\qquad z = c$

s.t. $\qquad\qquad s_t = s_{t-1} + i_t - q - w_t \qquad t = 1, 2, \ldots, n \qquad$ (1-1)

$\qquad\qquad s_t \leq c \qquad t = 1, 2, \ldots, n \qquad\qquad$ (1-2)

$\qquad\qquad s_n \geq s_o \qquad\qquad\qquad\qquad\qquad\qquad$ (1-3)

$\qquad\qquad s_t, w_t \geq 0 \qquad t = 1, 2, \ldots, n$

$\qquad\qquad c, q \geq 0$

The problem may be solved by any of the available standard linear programming codes. Problem size is such that solutions on a personal computer are not difficult. A simple specialized algorithm known as "sequent peak" that dates from the 1950s is available to solve the problem (see, e.g., Potter, 1977; Viessman and Welty, 1985).

Constraints (1) are mass balance constraints, really identities in a sense. They say that the ending storage for a particular month is equal to the storage at the end of the preceding month plus any inflow during the month less any release either to the water supply or to the stream itself. Constraints (2) prohibit any storage in excess of the storage capacity that is to be selected for the reservoir. Constraint (3) says that the storage at the end of the critical period must be at least as great as the unknown starting storage. This last constraint prevents "borrowing water" to artificially inflate the amount of water that can be delivered steadily throughout the course of the critical record.

The objective, slightly unusual for linear programming problems, consists of just a single variable, the unknown capacity to be chosen by the model. The general technique of minimizing the upper bound of some function or variable is known as "minimizing the maximum." The reader may ask, "Shouldn't the goal be to minimize costs?" Of course, minimum cost is the correct objective, but there is only one variable in the cost function, the reservoir capacity. Since the cost increases monotonically with capacity, it is sufficient to minimize capacity to minimize costs.

The linear programming formulation, which minimizes capacity subject to a yield requirement as well as its inverse, the maximization of yield given capacity, may exhibit an unexpected property when it is solved over a long sequence of inflows—if the sequence of flows includes a number of months of relatively high flows compared to the lowest flows. On occasion, positive spill quantities will be noted in some months—even when the end-of-period storages for those months are not at capacity. The notion that spill will only occur when storage bumps up against capacity is a correct idea—conceptually—in the sense that end-of-period storage equal to capacity can cause spill to occur. Nonetheless, the capacity to spill or the liberty to spill is more general.

That is, an LP solution, which includes spill in a month even though that month's end-of-period storage is less than capacity, is one of many alternate optimal solutions to the yield maximization or capacity minimization problem.

Such spill is disconcerting to observe but meaningless. The critical solution variables, the capacity or the yield, are not influenced by the presence or absence of such spill. In fact, an analyst can demonstrate this lack of influence by further calculation. Spill variables can be placed in the objective with a small weight and the weighted spill can be minimized. With suitable selection of a such a weight, all such spill variables can be pushed to zero without any impact on the key solution variables of the optimization—namely, yield and capacity. The operation of the reservoir would then look "logical," but the answer would not have changed. Thus, although the occurrence of such spills would seem to be an incorrect result, in reality, the analyst need not be concerned.

The problem can also be structured without the storage variable; partial sums replace the storage variable in the following formulation:

Minimize
$$z = c$$

s.t.
$$s_o + \sum_{t=1}^{k} i_t - \sum_{t=1}^{k} w_t - kq \leq c \qquad k = 1, 2, \ldots, n \qquad (1\text{-}4)$$

$$s_o + \sum_{t=1}^{k} i_t - \sum_{t=1}^{k} w_t - kq \geq 0 \qquad k = 1, 2, \ldots, n \qquad (1\text{-}5)$$

$$\sum_{t=1}^{n} w_t + nq \leq \sum_{t=1}^{n} i_t \qquad (1\text{-}6)$$

$$w_t \geq 0 \qquad t = 1, 2, \ldots, n$$

$$c, q \geq 0$$

This alternate formulation has n fewer variables and the same number of constraints as the earlier formulation. Instead of a storage variable, the storage at the end of k is represented by

$$s_o + \sum_{t=1}^{k} i_t - \sum_{t=1}^{k} w_t - kq$$

The storage at the end of k is made up of the initial storage plus all of the inflows up to and including the inflow in period k less all releases to the stream for want of further capacity less all planned releases to the water supply in all periods up to and including period k. The mass balance constraints (1-1) are not needed here because they are replaced by the partial sums. However, this formulation needs constraints (1-5), which require nonnegative end-of-period storages. These constraints were not needed in the first formulation because the nonnegativity requirements of the basic LP formulation enforced nonnegativity without formal constraints.

These two formulations both seek the smallest capacity needed to deliver a stable flow through the critical record without borrowing water. The formulations could be "turned around" with the capacity given and the largest sustainable flow sought as the objective. In this case, c is known and q is unknown.

Thus, instead of minimizing the maximum needed capacity, the formulation would maximize the sustainable release to water supply; that is, maximize q.

This largest sustainable release is often referred to as the "safe yield" of the reservoir. The term can be deceptive to the general public, however. It is "safe" to the water supply engineer because it was capable of delivery through the worst drought on record, but the engineer knows that worse droughts are entirely possible. The layperson assumes the term's conventional meaning, namely, without hazard. The delivery of this yield is not expected to be always safe. Hence, the term should probably not be used. Probably, "historical yield" could be used or "critical period yield" might be substituted so that people realize that the yield can only be sustained if the worst sequence of past inflows is repeated precisely.

The use of the critical period for sizing the reservoir or for gauging the historical yield represents the engineer's practical approach to the water supply. It is much like building a seaside cottage no nearer the sea than the highest recorded tides. In Florence, Italy, much of the famous statuary is presumably set on pedestals just above the height of the highest recorded flood waters of the Arno River. These practices reflect the judgment that future conditions are very unlikely to be worse than past conditions—or, if they are, sufficient capacity to adapt is present.

In fact, water supply engineers have developed means to assess the reliability of delivering a stable water supply quantity from a reservoir of stated capacity. The methodology, known as synthetic hydrology, is statistical in nature and consists of generating a number of artificial records of streamflows of lengths comparable to the existing record. The records are generated in such a way that all of the relevant statistical parameters of the original streamflow sequence are maintained in the new records. The critical period yield may be tested on these records using the capacity for which the yield was derived. If, for instance, the critical period yield can be sustained in 9 out of 10 of the records, the yield can be thought of as roughly 90% reliable or as having a 90% chance of being sustained in some future sequence of flows drawn from the same climactic regime. The reader interested in the methodology of synthetic hydrology is referred to Fiering and Jackson (1971). Hirsch (1979) provides an interesting investigation of the relations that can be built between synthetic hydrology and reliability.

Although synthetic sequences are not hard to generate, their use in assessing the reliability of a critical period yield has been questioned. Researchers have gradually discovered that the low flow sequences that make up the critical record typically may not be sufficiently well represented in the artificial records from synthetic hydrology. This fact, plus the unwieldy analysis needed to assess reliability, form the motivation for the new water supply reliability model, which is introduced in Chapter 7.

The concept of the critical period has allowed us to structure the basic capacity-minimizing or yield-maximizing linear programs for a water supply reservoir. Furthermore, the size of the problem we have needed to solve has been very small. The formulation using storage variables required only $2n + 2$ variables and $2n + 1$ constraints. Even if n were as large as 60 months, the problem would have only 122 variables and 121 constraints. However, as we noted earlier, the length and even the position of the critical period in the long record may shift as the yield is altered. When the yield is relatively small, the critical period is short. As the yield is pushed up, the length of the critical period increases and the position within the long record may move as well. In addition, the capacity required to deliver the specified yield also increases. The rate of increase of yield with capacity is constant as long as the length (number of months) and position of the critical period do not change. However, if either the length of the critical period is increased or its position shifts, then the rate of increase in yield with capacity will decline as capacity increases. The plot of yield versus capacity is known to water supply engineers as the storage-yield curve.

The fact that the critical period lengthens and shifts with an increasing water supply requirement and the fact that problem size is quite small for these programs—even for many years of record—suggests, as discussed earlier, that the original linear program should be run over a very long portion of the record. This long subrecord should include within it the critical months for the initial particular water supply requirement. In this way, the curve of yield versus capacity requires no recalibration of months of the critical period. That is, whatever the desired yield, the critical period will be within the total length of record being utilized for the calculation. Therefore, the number of months n should be much larger than the length of the critical period and the positioning of these months should encompass the initial critical period, at the least.

EVAPORATION ACCOUNTING ADDED TO THE BASIC MODEL

Reservoirs in hot, arid climates may suffer significant evaporation losses. Such losses, if not accounted for in the design of the reservoir, may impair the ability of the impoundment to perform its intended function(s). We offer here an approximate method to fold evaporation losses into the modeling of the water supply reservoir. New notation is needed.

We let

e_t = evaporation depth in linear units consistent with the units of the impoundment's surface area, known

a_t = average water surface area of the impoundment during month t

b = parameter that converts volume to billion gallons or whatever volumetric units are used for storage

The evaporation depth is a number that reflects temperature, humidity, and wind speeds for the month in question. Pan evaporation experiments over an extended period could provide such a number or could be used to establish a predictive equation that uses weather factors for calculating evaporation.

The area of the water surface is created by establishing a best-fit line of area against the water storage in the reservoir. The relation of area to storage volume is probably a function that displays both concavity and convexity depending on the range of volume level. A linear approximation of average area versus average volume is the order of the day here, so that we let

$$a_t = g + h\,\bar{s}_t$$

where g and h are the parameters of the best-fit line of area against the volume of water in storage, and $\bar{s}_t$ is the linear average storage volume in month t.

The volume of evaporation during month t then is given by

$$e_t\,a_t = e_t\,(g + h\,\bar{s}_t)$$

and this quantity is deducted from storage during month t.

The storage equation is now rewritten with this additional term subtracted from storage during month t:

$$s_t = s_{t-1} + i_t - q - w_t - e_t(g + h\bar{s}_t) \tag{1-7}$$

but $\bar{s}_t$ is approximated as the average of the month's beginning and ending storage:

$$\bar{s}_t = s_t/2 + s_{t-1}/2$$

The balance equation then becomes

$$s_t = s_{t-1} + i_t - q - w_t - e_t\left[g + (h/2)s_t + (h/2)s_{t-1}\right]$$

or

$$(1 + e_t h/2)s_t = (1 - e_t h/2)s_{t-1} + i_t - q - w_t - e_t g \qquad t = 2, \ldots, n \tag{1-8}$$

This last equation written in standard form replaces constraint (1-1) in the linear programming model that seeks either minimum capacity or maximum yield. If minimum capacity is sought, the values of i_t, q, and $e_t g$ remain on the right-hand side of the constraint since all these terms are constraints for the month t.[1]

---

[1]Although this procedure does incorporate evaporation into the design/operation of the single reservoir, a thoughtful consideration finds it a useful but approximate tool for this purpose. The evaporation depths are specific for each month of the year and reflect expected conditions for the month to which they are applied. In fact, the months of the critical period with their low flow levels probably represent hotter weather than the expected condition and hence greater amounts of evaporation than those expected normally. That is, the evaporation depths used in the model are probably not as great as they were during the critical period. Establishing evaporation depths reflective of those which occur during the months of the critical period is an interesting challenge.

The topic of evaporation accounting should not be left without noting that a differential equation model could be built for the storage volume equation. Such a model would make the evaporation rate depend on the instantaneous area or storage volume. The differential equation might read

$$\left.\frac{ds}{d\theta}\right)_t = i_t - q - w_t - e_t g - e_t hs \tag{1-9}$$

where θ is the time in month t, and where all terms but the last can be treated as a constant for purpose of solving this first-order linear differential equation. That is,

$$\left.\frac{ds}{d\theta}\right)_t = b_t - k_t s \tag{1-10}$$

where

$$b_t = i_t - q - w_t - e_t g$$

and

$$k_t = e_t h$$

The integrated form of this equation is

$$s_t = (s_{t-1} - b_t/k_t)e^{-k_t\theta} + b_t/k_t \tag{1-11}$$

so that at the end of the period, when $\theta = 1$, the storage is given by

$$s_t = (s_{t-1} - b_t/k_t)\,e^{-k_t} + b_t/k_t$$

or

$$s_t = (e^{-k_t})s_{t-1} + (b_t/k_t)(1 - e^{-k_t}) \tag{1-12}$$

Returning to the form for the linear programming constraint, we have

$$s_t = (e^{-k_t})s_{t-1} + k_t' i_t - k_t' q - k_t' w_t - k_t' g$$

where

$$k_t' = (1 - e^{-k_t})/k_t$$

This form, if it is preferred to the more approximate form based on average storage in the interval, could also replace constraint (1-1) in the linear programming model.

WATER SUPPLY AND FLOOD CONTROL

Flood control, in contrast to water supply, requires free volume in the reservoir to catch and store floods. The captured flood waters are gradually released at rates that do not cause damage to downstream users or at rates that minimize the damage that might be caused. We consider next a reservoir that provides both the water supply and flood control functions. Thus, a portion of the reservoir is dedicated to water supply storage and another portion to emptiness. Methods of assessing the volume needed for flood storage in each month or throuhgout the year are designed to minimize expected damages or to achieve a sufficiently small probability of spillway use or of overtopping the reservoir.

One method to calculate flood storage pools involves translation of flood pool volumes into the maximum discharge rates that would be necessary. These discharge rates yield stage levels (height) of the stream that in turn determine damages. The portion of the reservoir devoted to the flood pool and the portion devoted to water supply are chosen so that the sum of water supply benefits and flood damages prevented are as large as possible. The goal in this section is to show how to determine the largest sustainable yield that can be achieved given (1) a historical record of inflows, (2) a total storage capacity c for the reservoir, and (3) 12 flood storage volumes or flood pools, one for each month of the year. The difference between this model for water supply and the basic model is that in the basic model, the maximum capacity available for water storage was the same month to month. In this model, storage available for water supply at the end of each month is reduced from the reservoir capacity by the amount of the flood storage volume. These numbers for flood pool volumes, which may be specific by month or constant throught the year, are indicated as

v_k = flood pool volume to be maintained in the reservoir during month k of the year, known, in billions of gallons

These numbers can now be incorporated into the basic reservoir model in which the reservoir capacity, c, is already known. Using the notation we introduced at the outset, we can seek the largest steady monthly supply that can be provided by the reservoir through some critical period or long record of stream flow.

The model of maximizing sustainable yield may be stated as

Maximize $z = q$
s.t. $s_t = s_{t-1} - q - w_t + i_t$ $t = 1, 2, \ldots, n$ (1-14)

$s_t \leq c - v_k$ $t = 1, 2, \ldots, n;$ (1-15)
$k = t - 12[(t - 1)/12]$

$s_n \geq s_o$ (1-16)
$s_t, w_t \geq 0$ $t = 1, 2, \ldots, n$
$q \geq 0$

and

$$[u] = \text{integer part of } u$$

Once again, the constraints (1-14) represent the mass balance statement. Constraints (1-15) limit month k end-of-period storages to $c - v_k$, and constraint (1-16) prevents borrowing water from initial storage. Constraints (1-15) are written so that in each recurrence of month k in successive years, the appropriate upper limit on water storage is always enforced. For instance, suppose $t = 26$, and we are referring to the second month of the third year. Then $[(t - 1)/12] = 2$. The value of $k = 26 - 12 \times 2 = 2$. Hence,

$$s_{26} \leq c - v_2$$

Since v_k is the volume needed during month k, the model should actually be modified so that the storage through the month is less than $c - v_k$. This might be approached by constraining the average storage in the month less than that value. This model could also possibly be extended to a multiple reservoir system that serves both flood control and water supply functions. (Later, the basic water supply model is extended to the multiple-reservoir water supply system.)

This model is for monthly water supply operation given a reservoir of a fixed size with specified monthly flood control volumes. The model does not guide a reservoir manager in operation during the hours or days of a flood. The reader interested in the question of operation during floods can consult articles listed in the "References in Operation for Flood Control."

WATER SUPPLY WITH MONTHLY VARYING RELEASE

When the release varies by month, the objective of the linear program can no longer be simply to maximize the steady month-to-month release. Here then we are in the situation of assessing how much water can be delivered from a reservoir of known capacity when the quantity to be delivered is specific for each month of the year. There is a January release to water supply, a

February release, a March release, and so on. We do not know what these numbers are, only that they are the same numbers year to year. Furthermore, we can deduce, since they are the same year to year, that the January release is always the same proportion of the total of the 12 monthly releases; the February release is a constant fraction of the sum of the 12 releases; and so on. The number that is singular, that is, the same year to year, although still unknown, and that we can optimize, is the sum of the 12 monthly releases, namely, the annual yield.

We introduce the following new notation to supplement earlier notation:

Q_A = annual release to water supply, unknown
β_k = fraction of the annual yield that is released in month k (k = 1, 2, . . . , 12), known

The β_k, positive fractions that sum to one, would be expected to be largest in the drier summer and fall months and least in the wetter winter and spring months. The record of inflows over which the model is solved should be the multiple years that encompass all possible critical periods. The maximum annual yield model can now be structured as

Maximize $\quad z = Q_A$

s.t. $\qquad s_t = s_{t-1} + i_t - w_t - \beta_k Q_A \qquad t = 1, 2, \ldots, n; \qquad$ (1-17)
$\qquad\qquad\qquad\qquad\qquad\qquad\qquad\quad k = t - 12[(t - 1)/12]$

$\qquad s_t \leq c \qquad t = 1, 2, \ldots, n \qquad\qquad\qquad$ (1-18)

$\qquad s_n \geq s_o \qquad\qquad\qquad\qquad\qquad\qquad\qquad\qquad$ (1-19)
$\qquad s_t, w_t \geq 0 \qquad t = 1, 2, \ldots, n$
$\qquad Q_A \geq 0$
$\qquad [u]$ = integer part of u

Constraints (1-17) are the mass balance equations modified to reflect different releases in each of the 12 months. To illustrate, suppose $t = 26$. Then $k = 2$ and

$$s_{26} = s_{25} + i_{26} - w_{26} - \beta_2 Q_A$$

That is, the release $\beta_2 Q_A$ is taken from the reservoir contents toward the February requirement. Constraints (1-18) limit storage to the reservoir's capacity, and constraint (1-19) prevents borrowing water.

As in the basic model, a storage-yield curve may be constructed. The annual water delivery Q_A will be found to increase in successively lower-sloped linear segments as the reservoir capacity c is allowed to increase. That is, there are declining returns to scale. Each linear segment corresponds to either an expanded-in-length or shifted-in-position critical period.

WATER SUPPLY OPERATING RULES AND SIMULATION

The model that maximizes the steady month-to-month yield, given a capacity c, follows an *operating rule,* albeit a simple operating rule. An operating rule is an equation or a chart or a look-up table that specifies the amount to be released for various purposes as a function of system states and parameters. Such rules become most necessary when the reservoir is in a state of "stress," that is, when demands are becoming difficult to meet without threatening future supplies. Most often, those system characteristics that determine release are state variables such as reservoir storage at the end of the previous period, or stream flows, such as the flow in the immediately preceding month and the flow projected for the next month. Release may be composed of linear or nonlinear functions of these separate descriptors or may be a function of the sum of the descriptors or could even take other forms. We will provide further discussion of possible operating rules later, but for now, we return to a consideration of the operating rule assumed by the linear programming reservoir model.

The LP model assumes that release is equal to the water supply level q when the sum of the previous reservoir storage, and this month's inflow are within the following range. On the lower end of the range, the sum is q. If storage plus inflow sum to q or greater, q is released. However, if previous storage plus this month's inflow less q exceeds the capacity of the reservoir, then spill w_t must take place.[2] The amount of the spill is the difference between storage plus inflow less q and less the capacity c. If the storage plus inflow are less than q, clearly q cannot be released as the formalism of release would lead to a negative storage.

Since q is the largest possible steady release as determined by the LP that can be achieved without a negative reservoir storage with a reservoir capacity of c over a critical record, then never, over that record, will there be a need to release less than q. If, however, a quantity larger than q were attempted steadily over the critical record, at some moment storage would go negative. Hence, if a quantity larger than q were attempted, nevertheless, the reservoir operator would be forced at times during the critical record to decrease the release to water supply. It would not be physically possible to do otherwise. The rule implied by the linear program, however, never needs to make such a choice because the q chosen is *always* achievable over the record being investigated.

Let us assume that a release to water supply is chosen that is greater than q; say, r, a number that is not always physically achievable through the record, given the capacity of the reservoir. Now a rule does need to be created for the situation in which the previous storage plus current inflow is less then the desired water supply release. The simplified rule we will use first, if the desired

---

[2]This "spill" is a planned wasting of water to the stream through gates or release ports rather than water coursing through the spillway.

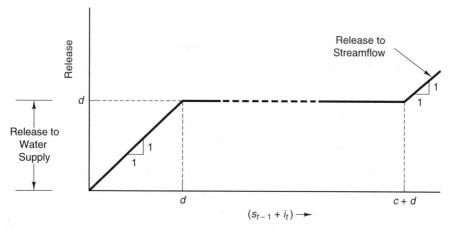

FIGURE 1-1   Water supply spill-release rule.

supply is not available, is to release the entire contents of the reservoir, storage plus inflow, toward water supply.

To recapitulate, in this new situation in which a water supply release of r is desired, the following responses occur. If storage plus this month's inflow are less than r, release the storage plus inflow. If storage plus inflow exceed r and the combined storage and inflow less r is less than capacity, release r precisely. If storage plus inflow less r exceed capacity, release the difference or amount of exceedance as spill to the stream, in addition to r. The previously discussed relations are summarized in the following equations and in Figure 1-1.

New notation is needed:

x_t = release to water supply in month t
r = desired water supply release

The release to water supply in month t is

$$x_t = s_{t-1} + i_t \qquad s_{t-1} + i_t < r \tag{1-20}$$

$$x_t = r \qquad s_{t-1} + i_t \geq r$$

The spill release to stream w_t is

$$w_t = 0 \qquad s_{t-1} + i_t \leq c \tag{1-21}$$

$$w_t = s_{t-1} + i_t - r - c \qquad s_{t-1} + i_t > c$$

Finally, the reservoir contents are given by

$$s_t = s_{t-1} + i_t - x_t - w_t \qquad\qquad (1\text{-}22)$$

where x_t and w_t are determined by the previous equations. Alternatively, reservoir contents could also be written as

$$s_t = 0 \qquad s_{t-1} + i_t < r$$
$$s_t = s_{t-1} + i_t - r \qquad r \le s_{t-1} + i_t < c + r$$
$$s_t = c \qquad s_{t-1} + i_t - r \ge c$$

The reservoir release rule and its consequences, (1-20), (1-21), and (1-22), constitute a pattern to guide monthly operation of a reservoir over many years of record. The choice of a water supply release quantity obviously influences the reservoir contents through time. A release set equal to q, the maximum sustainable steady release, will lead to a reservoir that, though it may empty at times, never needs to cut back its release for lack of storage and inflow if operation is performed over the historical record. A release smaller than q will lead to a reservoir that never empties over the record and that has a higher average storage and that spills water to downstream flow more often. A release larger than q leads to a reservoir that has to cut back its release to some extent on some occasions, that hits bottom more frequently than when the steady release is q, and that has a lower average storage, spilling less frequently than when the steady release is q. The larger that the steady release is made, as long as it is above q, the more frequent and the greater in extent are release cutbacks. In addition, average storage decreases with increasing steady release, as does the frequency of release to downstream flow.

It is a relatively easy step to vary the monthly need, to make it r_k for the kth month of the year, where the r_k repeat every 12 months that elapse. Nothing new is added conceptually to the modeling steps.

The process of adding inflows and subtracting releases month by month is a form of *simulation*. Simulation means repeated operation in an artificial as opposed to real environment. In contrast to optimization, which locates some optimal policy or design, simulation tests alternative policies and designs and compares them in as many perspectives as possible. Each repetition of a simulation alters some facet of the environment, such as the steady release or the capacity of the reservoir or perhaps the rate of release in the situation in which the full water supply quantity cannot be delivered.

These repeated tests can lead to a number of useful plots. For instance, for a given capacity, the frequency of needing to cut back from the otherwise steady delivery may be plotted against the target delivery rate. The same variables can be plotted at different capacity levels. Alternatively, for a given steady release, the frequency of cutbacks might be plotted against capacity. Reservoir simulation can be a useful technique when the system is simple, for example, one reservoir, and the functions few, for example, water supply and streamflow maintenance. When the system is more complex, analysis by simulation requires "art" to tease relevant information from the system in a useful form.

In a moment, we will describe other operating rules that can be tested by simulation, but for now, we shall list in more detail the measures of performance that can be tabulated. The list is not complete, but meant to be suggestive of the range of possibilities. The data for each statistic are compiled over a long record of operation that may include multiple low-flow periods. So far, we have only simulated a reservoir delivering a constant release, but these statistics could be compiled for release targets that vary with the month of the year.

1. the frequency of shortage, that is, the number of months in which the release is less than the target for the month
2. the largest shortage, as calculated by the amount less than the target for the month. This would be referred to as the "depth" of shortage
3. the largest fractional shortage, that is, the largest of the ratios of shortage to target over all months
4. the average summertime shortage
5. average storage in the reservoir
6. the frequency of spills (water wastage)
7. the largest "spill" recorded
8. the maximum number of consecutive months that shortages were incurred (the maximum length of a drought)
9. the number of times that two consecutive months of shortage were recorded—where each "string" of shortages is counted separately

The operating rule we have described is the rule implied by the linear programming model with the additional feature that the release during any month of shortage is precisely the quantity of water that is available. Despite the fact that the rule very nearly corresponds to the linear programming rule, it is, nevertheless, very simplistic. If a reservoir were "under stress," if the monthly target could be seen not to be sustainable over a potential drought period, this rule could lead to a severe shortage. By never cutting back, by constantly delivering the target despite the concern of a possible shortage, the reservoir could enter a condition of storage and future inflow in which, for a particular month or months, a dramatic reduction in draft may be needed. If, on the other hand, when reservoir contents entered some range of concern, and the release over succesive months was diminished from its target value through rationing or other reductions, then the month or months of severe reduction might be avoided. That is, some months of small shortfalls (and consequent small economic losses) could replace a brief period of severe shortfall (and potentially large losses).

Rules that might effectively guide the operation of a reservoir under stress could look like Figure 1-2, but this figure is only an example.

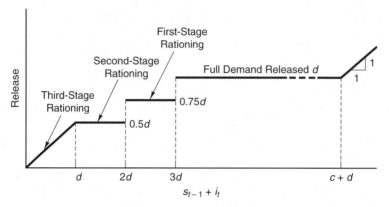

FIGURE 1-2   A water supply rationing rule.

Water supply operation under a rule such as that illustrated in Figure 1-2 would proceed as follows. When the sum of the previous end-of-month storage and this month's projected inflow exceed three month's supply, the full demand for the current month is released. When the sum of storage and projected inflow is less than three month's supply but greater than two month's supply, first-phase rationing is declared; decreases in use may be promoted by both requests for cooperative behavior and economic coercion. The rule shown suggests that demand may be reduced by 25% with first-phase efforts. The mix of requests and penalties, however, determine actual reductions.

When the sum of storage and projected inflow falls in the range between two and one month's need, second-phase rationing may be instituted with its mix of requests for cooperation and economic penalties. The picture suggests a 50% reduction, but, again, the mix and nature of the incentives determine actual reductions. When the sum of storage and projected inflow fall below one month's needs, release of this sum is suggested by the rule. Of course, more phases of rationing are quite possible. Further, the trigger volumes that initiate the phases of rationing are parameters that may be manipulated in experiments to determine efficient levels for these values.

As before, when storage plus projected inflow less the full demand exceed the capacity of the reservoir, planned spill will take place. All these relations are summarized by equations that parallel equations (1-20), (1-21), and (1-22). These rules would be used to operate a reservoir over a long record of inflows. Statistics would be compiled that indicate such measures as

1. the frequency of phase 1 rationing
2. the frequency of phase 2 rationing
3. the frequency of phase 3 rationing
4. the frequency of phases 1 and 2 together
5. the frequency of months with any phase (1, 2, or 3) of rationing

6. the maximum number of months in succession with any level of rationing, or with phase 1 rationing, or with phase 1 or phase 2, and so on

7. the largest depth of drought

and so on.

It should be clear that the larger the first trigger volume, the more frequent phase 1 rationing will be and the less frequent phase 2 and phase 3 rationing will be. The proper setting of all of these trigger volumes requires artful manipulation of their levels in multiple simulation runs.

Simulation is a very popular method of water resources analysis, and a number of agencies and organizations have prepared their own simulation codes. These include the U.S. Army Corps of Engineers, the Colorado Water Resources Institute at Colorado State University, the Texas Water Development Board, and the Center for Advanced Decision Support for Water and Environment Systems (University of Colorado). The reader interested in water resources simulation can consult the literature in the "References in Water Resources Simulation."

BIBLIOGRAPHY

The Single Water Supply Reservoir

Dorfman, R. 1965. "Formal Models in the Design of Water Resource Systems." *Water Resources Research,* Vol. 1, No. 3, pp. 329–336.

Fiering, M. 1967. *Synthetic Hydrology.* Harvard University Press, Cambridge, Mass.

Fiering, M., and B. Jackson. 1971. *Synthetic Streamflows,* Water Resources Monograph No. 1. American Geophysical Union, Washington, D.C.

Hashimoto, T., J. Stedinger, and D. Loucks. 1982. "Reliability, Resilience, and Vulnerability Criteria for Water Resource System Performance Evaluation." *Water Resources Research,* Vol. 18, pp. 14–20.

Hirsch, R. 1979. "Synthetic Hydrology and Water Supply Reliability." *Water Resources Research,* Vol. 15, No. 6, pp. 1603–1615.

Howe, C., and F. Linaweaver. 1967. "The Impact of Price on Residential Water Demand and Its Relation to System Design and Price Structure." *Water Resources Research,* Vol. 3, pp. 12–32.

Moy, W.-S., J. Cohon, and C. ReVelle. 1986. "A Programming Model for Analysis of the Reliability, Resilience, and Vulnerability of a Water Supply Reservoir." *Water Resources Research,* Vol. 22, pp. 489–498.

Potter, K. 1977. "Sequent Peak Procedure: Minimum Reservoir Capacity Subject to Constraint on Final Storage." *Water Resources Bulletin,* Vol. 13, No. 3, pp. 521–528.

Shih, J.-S., and C. ReVelle. 1994. "Water Supply Operations During Drought: Continuous Hedging Rule." *Journal of Water Resources Planning and Management (ASCE),* Vol. 120, No. 5, pp. 613–629.

Shih, J.-S., and C. ReVelle 1995. "Water Supply Operations During Drought: A Discrete Hedging Rule." *European Journal of Operational Research,* Vol. 82, pp. 163–175.

Viessman, W., and C. Welty. 1985. *Water Management: Technology and Institutions.* Harper and Row, New York.

Vogel, R., and R. Bolognese. 1965. "Storage-Reliability-Resilience-Yield Relations for Over-Year Water Supply Systems." *Water Resources Research,* Vol. 31, No. 3, pp. 645–654.

Yeh, W. 1985. "Reservoir Management and Operations Models: A State-of-the-Art Review." *Water Resources Research,* Vol. 21, No. 12, pp. 1797–1818.

Operation for Flood Control

Karbowski, A. 1993. "Optimal Flood Control in Multireservoir Cascade Systems with Deterministic Inflow Forecasts." *Water Resources Management,* Vol. 7, pp. 207–223.

Kelman, J., and J. M. Damazio. 1989. "The Determination of Flood Control Volumes." *Water Resources Research,* Vol. 25, No. 3, pp. 337–344.

Marien, J. L., J. M. Damazio, and F. S. Costa. 1984. "Building Flood Control Rule Curves for Multipurpose Multireservoir Systems Using Controllability Conditions." *Water Resources Research,* Vol. 30, No. 4, pp. 1135–1144.

Pytlak, R., and K. Malinowski. 1989. "Optimal Scheduling of Reservoir Releases During Flood: Deterministic Optimization Problem, Part 1, Procedure." *Journal of Optimization Theory and Applications,* Vol. 61, No. 3, pp. 409–449.

Unver, O., and L. Mays. 1990. "Model for Real-Time Optimal Flood Control Operation of a Reservoir System." *Water Resources Management,* Vol. 4, pp. 21–46.

Wasimi, S., and P. K. Kitanidis. 1983. "Real-Time Forecasting and Daily Operation of a Multireservoir System During Floods by Linear Quadratic Gaussian Control." *Water Resources Management,* Vol. 19, No. 6, pp. 1511–1522.

Water Resources Simulation

Browder, L. E. 1978. "RESOP-II Reservoir Operating and Quality Routing Program, Program Documentation and User's Manual," UM-20. Texas Department of Water Resources (now renamed Texas Water Development Board), Austin.

Feldman, A. D. 1981. "HEC Models for Water Resources System Simulation: Theory and Experience." *Advances in Hydroscience* (V. T. Chow, Ed.), Vol. 12. Academic Press, New York.

Kuczera, G., and G. Diment. 1988. "General Water Supply System Simulation Model: Wasp." *Journal of Water Resources Planning and Management (ASCE),* Vol. 114, No. 4.

Labadie, J. W., A. M. Pineda, and D.A. Bode. 1984. "Network Analysis of Raw Supplies Under Complex Water Rights and Exchanges: Documentation for Program MODSIM3." Colorado Water Resources Institute, Fort Collins, Col.

Sigvaldason, O. T. 1976. "A Simulation Model for Operating a Multipurpose Multireservoir System." *Water Resources Research,* Vol. 12, No. 2.

Strzepek, K. M., L. A. Garcia, and T. M. Over. 1989. *MITSIM 2.1 River Basin Simulation Model, User Manual.* Center for Advanced Decision Support for Water and Environmental Systems, University of Colorado, Boulder.

U.S. Army Corps of Engineers, Hydrologic Engineering Center. 1982. *HEC-5 Simulation of Flood Control and Conservation Systems, User's Manual.* Davis, Calif.

U.S. Army Corps of Engineers, Hydrologic Engineering Center. 1989. *HEC-5 Simulation of Flood Control and Conservation Systems, Exhibit 8, Input Description.* Davis, Calif.

CHAPTER 2

THE WATER SUPPLY FUNCTION: MULTIPLE RESERVOIRS

INTRODUCTION

In this section, we extend the single water supply reservoir model to a system of reservoirs in parallel. By "in parallel," we mean that each reservoir is on a parallel river above any point of juncture with the other rivers in the system. Reservoirs on the same stream, that is, in series, do not require special modeling unless times of flow between the reservoirs are more than a few days or a week. Such reservoirs in series on a single stream may generally be treated as a single reservoir for water supply operational purposes.

The basic issue we will examine for operation of reservoirs on parallel rivers is how much yield, either a steady through-the-year monthly yield or an annual yield with individual yields that vary with month of the year can be obtained from the system of reservoirs. One overly simple and overly conservative answer to this question is that the yield from the reservoir system is the sum of historic or critical period yields of the individual reservoirs. This specific answer presumes that each reservoir will continue to act separately as though it were not a part of a larger system. In such a system, the only gain from the addition of a new reservoir is its individual historic yield.

In fact, considerable yield advantage may be obtained from operating a system of reservoirs *as a system*. The differences in local watershed climates often produce seasonal differences in streamflows. These seasonal differences can be translated into different monthly contributions from the individual reservoirs, contributions that may vary by month across the reservoirs but sum to the desired total for each month.

System operation is not yet a finished science. There is no single optimal way to operate a system of reservoirs for water supply—especially in the face of an unknown future. There are not only multiple approaches to scientific operation. There is also the school of "experience," widely used still as we begin to transist into an era of scientific reservoir management. The school of "experience" draws on the accumulated years of operating wisdom of reservoir system managers. It is not to be dismissed. Nonetheless, the mathematical approaches to be discussed next are designed to supplement expert knowledge and experience, and, also, to provide rules against which expert judgment can be tested.

We will structure the problem of reservoirs in parallel as one of either maximizing the single yield that can be sustained month to month or maximizing the annual yield. The reservoirs will be assumed to be in place with storage capacities at prespecified levels. The numbers we obtain represent a possible upper bound on the system yield, the best yield that careful management practice can provide, given a repetition of the worst system flow conditions of the past. That is, once again, we will seek to operate through a critical period, but, in this case, a critical period that has wider spatial and probably temporal boundaries.

What is the justification of seeking the maximum system yield? The justification of this objective is that demand almost always grows with time. The value of the maximum yield coupled with a projection of demand growth tells us how long the physical system will be able to provide sufficient water before new sources of water supply or water conservation efforts will be needed. Of course, the cost of water conservation should be balanced against the full cost (economic and environmental) of additional water supplies.

There is an obverse or inverted form of this problem statement, just as in the case of the single reservoir. For the single reservoir, we can minimize capacity—as a surrogate for cost—given a specified yield. Or we can maximize yield subject to a particular level of capacity. Both problems are straightforward. For the reservoir system, the inverted problem form to the one stated before is to assume the reservoirs are not in place and to determine the individual capacities that minimize the cost of the reservoir system given a required water supply yield. Unfortunately, this inverted problem is not as easy to solve as the problem that maximizes yield given capacities. One version of the inverted or cost-minimizing system problem will be structured after we have treated the yield-maximizing problem.

To begin, we will structure the problem of maximizing yield given reservoirs in place. We will assume that a single reservoir is on each of three nonintersecting streams and that a contribution toward water supply can be drawn from each reservoir.

MAXIMIZING YIELD WITH RESERVOIRS IN PLACE

Maximum system yield may be approached in two different ways or with two different philosophies. In one way, contributions that are the same from year to year are calculated for each reservoir for each month of the year under the assumption that a sufficient portion of the "gain" from joint operation can be captured by such a simple monthly allocation. That is, on a given month of the year, we calculate for each reservoir a specific contribution toward system yield, a contribution that does not vary from year to year, except under extraordinary conditions. These contributions can be calculated by solution of optimization models.

In a second way, achieving the maximum system yield is approached as a problem in which the monthly contribution of each reservoir is variable and is reflective not only of its status of fullness and anticipated inflows, but also reflective of the storage and anticipated inflows of the other reservoirs in the system. Calculating such variable contributions may be approached as an optimization problem or as a simulation problem. Both methodologies are presented here.

We begin with the problem of determining for a stated system yield the monthly reservoir contributions, contributions that do not vary from year to year. We proceed next to the situation in which contributions toward the common target reflect current conditions in the system.

Monthly Contributions Do Not Vary Year to Year

The linear programming model presented here for the maximization of system yield is a simplified modification of the model of Palmer and colleagues (1982). For simplicity, we assume there are three reservoirs in the system. The following notation is needed:

c_1, c_2, c_3 = capacities of each of three reservoirs

i_{1t}, i_{2t}, i_{3t} = stream flow sequences for month t on each of the three streams. The record length is given as n months. The record encompasses the worst overlapping flow sequences, that is, those that define the system yield

q_{1k}, q_{2k}, q_{3k} = release toward water supply from each of the three reservoirs for month k of the year

w_{1t}, w_{2t}, w_{3t} = unknown release to streamflow from each of the three reservoirs in month t

s_{1t}, s_{2t}, s_{3t} = end-of-month storages in each of the three reservoirs

q_M = largest cumulative draft that can be achieved month after month from the three reservoirs toward the system water supply, unknown

The linear program can be stated as

$$\text{Maximize} \quad z = q_M$$

$$\text{s.t.} \quad s_{1t} = s_{1t-1} + i_{1t} - w_{1t} - q_{1k} \quad \begin{aligned} & t = 1, 2, \ldots, n; \\ & k = t - 12[(t-1)/12] \end{aligned} \quad (2\text{-}1)$$

$$s_{1t} \leq c_1 \qquad t = 1, 2, \ldots, n \tag{2-2}$$

$$s_{2t} = s_{2t-1} + i_{2t} - w_{2t} - q_{2k} \qquad \begin{aligned} t &= 1, 2, \ldots, n; \\ k &= t - 12[(t-1)/12] \end{aligned} \tag{2-3}$$

$$s_{2t} \leq c_2 \qquad t = 1, 2, \ldots, n \tag{2-4}$$

$$s_{3t} = s_{3t-1} + i_{3t} - w_{3t} - q_{3k} \qquad \begin{aligned} t &= 1, 2, \ldots, n; \\ k &= t - 12[(t-1)/12] \end{aligned} \tag{2-5}$$

$$s_{3t} \leq c_3 \qquad t = 1, 2, \ldots, n \tag{2-6}$$

$$s_{in} \geq s_{io} \qquad i = 1, 2, 3 \tag{2-7}$$

$$q_{1k} + q_{2k} + q_{3k} = q_M \qquad k = 1, 2, \ldots, 12 \tag{2-8}$$

Of course, all variables are constrained greater than or equal to zero.

Equations (2-1), (2-3), and (2-5) are the mass balance equations for each of the three reservoirs. Constraints (2-2), (2-4), and (2-6) limit storages to the known reservoir capacities. Constraint (2-7) compares to constraint (1-19), which is reservoir-specific, requiring that the ending storage in each reservoir exceeds the beginning and unknown storage in each reservoir. Constraint (2-8) requires that the sum of releases in month k (e.g., February) equal the steady monthly yield q_M.

The crucial answers to this problem, as in the single-reservoir model, are not the reservoir storages through time or the spill quantities. The crucial answers from this program are the 36 q's, three for each month of the year. These are the allocations toward the steady monthly system yield to be drawn from each of the three reservoirs for each of the 12 months of the year. For any month, the three reservoir drafts sum to system yield, but for any particular reservoir, the numbers may change by month through the year. In some months, one of the reservoirs may contribute a large share of the system yield, whereas in others, the same reservoir may contribute a small share. This contrasts to the situation in which the three reservoirs are operated independently of one another and the system yield (likely a smaller number) is obtained by merging the individual yields of the reservoirs.

The inconsistency of release from any particular reservoir may be disconcerting and may be eliminated by further programming. We might proceed by maximizing system yield with constraints for each reservoir i that limit the month-to-month change in that reservoir's contribution to some level.

That is, maximize q subject to all of the original constraints and

$$q_{ik} - q_{i,k+1} \leq \Delta \qquad k = 1, 2, \ldots, 11$$

$$q_{i,k+1} - q_{ik} \leq \Delta \qquad k = 1, 2, \ldots, 11$$

These two constraints limit the increase or decrease to Δ. The reduction in system yield is likely to be small, and as Δ is allowed to increase, smaller still.

The basic multiple-reservoir model can also be structured for the situation in which the system requirement cycles with month or season. In this case, we seek to maximize the annual yield (Q_A) rather than monthly yield. Specifically, we

$$\text{Maximize} \quad z = Q_A$$

subject to constraints (2-1) to (2-7) inclusive. Equations (2-8) are now modified to

$$q_{1k} + q_{2k} + q_{3k} = \beta_k Q_A \quad k = 1, 2, \ldots, 12 \qquad (2\text{-}9)$$

where β_k is the portion of annual yield that is attributed to month k. Once again, the crucial result is not only the annual yield, but the monthly contribution from each reservoir toward the monthly requirement $\beta_k Q_A$.

The rules we have formulated thus far for the operation of the parallel reservoirs may be termed "zero-order" rules. They depend on no information that is current at the time that the individual reservoir releases are made. Another term applied to describe these rules is "once-and-for-all," in the sense that individual releases are specified in advance and are not adjusted for any current conditions.

Other classes or orders of rules may be formulated that do utilize current information. For instance, a first-order rule might specify that the release from a particular reservoir is linearly proportional to storage or inflow or the sum of these for that reservoir. Such rules as these might be termed "here-and-now" rules in the sense that they adjust to or are attuned to current conditions. Another descriptive term for such operating guidelines is "real time" operating rules. We will suggest several approaches to yield maximization in real time. These approaches will include simulation and optimization in real time.

Monthly Contributions Vary Year to Year

The Simulation Approach to Yield Maximization

Heuristic Operating Rules and Yield Maximization   One way to deal with the issue of yield maximization using rules that adjust reservoir contributions to current conditions is to specify the precise rules in advance and then test their performance in providing a particular yield level. The specification of these rules clearly is a heuristic—since the best self-adjusting rules cannot be known a priori.

Suppose the rules are specified. Suppose further that a value for maximum yield has been estimated. The estimate of maximum yield and the rules can be tested together using a long record that includes the worst system droughts experienced over the years of flow measurement. The record should begin in years of ordinary rather than drought flows so that the reservoirs begin their years of crucial operation with realistic opening storage conditions. When the long record opens, the reservoirs will be assumed full, a reasonable assumption during years of ordinary flow.

The testing, of course, is by a simulation, much like that described for the single water supply reservoir. We will offer a number of alternative rules for allocating releases amongst reservoirs. We need to define six additional parameters to those defined in Chapter 1 in order to structure the various rules:

ip_{it} = projected flow for month t into reservoir i, known
ir_i = anticipated flow into reservoir i during the remainder of the draw-down-refill cycle, known
α_{it} = unknown proportion of month t's demand allocated to reservoir i
x_{it} = calculated contribution (volume) of reservoir i in month t, unknown
T = estimate of system yield to be tested
v_{it} = free volume in reservoir i at the end of period t

The projected flow number may be based on a correlation of this month's streamflow with last month's, or it may be based on snowpack in a mountainous area, or it may be estimated by some other means. This is a projected as opposed to actual flow for month t. Both actual and projected flows may be carried for each month. The projections, if they are carried, are used to calculate system operation (the month's calculated releases) at the beginning of a month. The actual flows are always carried and are used to update the reservoir contents after the month's calculated releases have been made.

The first of the rules we discuss is Clark's "space rule," which Maas and colleagues (1962) describe as

Apportion the required release the reservoirs in such a way that, after water has been withdrawn, the ratio of the space available in each reservoir to that in all reservoirs equals, in so far as possible, the ratio of the predicted flow into each reservoir during the remainder of the drawdown-refill cycle to that in all reservoirs.

The free volume in reservoir i projected for the end of period t is given by

$$v_{it} = c_i - (s_{it-1} + ip_{it} - x_{it})$$

where the term in parentheses is the storage projected for reservoir i at the end of the period. The ratio of free volume in reservoir i to the total free volume in the system is

$$\frac{[c_i - (s_{it-1} + ip_{it} - x_{it})]}{\sum_{i=1}^{3} [c_i - (s_{it-1} + ip_{it} - x_{it})]}$$

and

$$\alpha_{it} = \frac{ir_i}{\sum\limits_{k=1}^{3} ir_k}$$

equals the preceding ratio.
    Therefore,

$$\frac{[c_i - (s_{it-1} + ip_{it} - x_{it})]}{\sum\limits_{i=1}^{3} [c_i - (s_{it-1} + ip_{it} - x_{it})]} = \alpha_{it}$$

but the sum of releases from the three reservoirs equals the system yield:

$$\sum_{i=1}^{3} x_{it} = T$$

so that the free-volume-ratio statement becomes

$$\frac{[c_i - (s_{it-1} + ip_{it} - x_{it})]}{\left[\sum\limits_{i=1}^{3} [c_i - (s_{it-1} + ip_{it})] + T\right]} = \alpha_{it}$$

We can set the summation terms in the denominator equal to a constant because all of its terms are known.

$$\sum_{i=1}^{3} (c_i - s_{it-1} + ip_{it}) = N$$

This gives us an equation we can solve for x_{it}, namely,

$$c_i - (s_{it-1} + ip_{it} - x_{it}) = \alpha_{it}(N + T)$$

giving us the contribution from reservoir i in the month t:

(A) $x_{it} = \alpha_{it}(N + T) - c_i + s_{it-1} + ip_{it}$ $i = 1, 2, \ldots, 3$

This rule, the space rule, is designed to minimize the water spilled in the system during the remainder of the drawdown-refill cycle. It does not, however, *guarantee* minimum system spill. In addition, the rule does not give a value for the joint historic yield, only an allocation formula for relative contributions.

Further, an estimate, T of the joint historic yield needs to be provided to apply the allocation formula.

Another rule for the monthly contributions of the individual reservoirs toward the historic yield for the system is

$$(B) \qquad x_{it} = \left[\frac{s_{it-1}}{\left(\sum\limits_{i=1}^{3} s_{it-1} \right)} \right] * T \qquad i = 1, 2, 3$$

Rule B states that the allocation from reservoir i is in direct proportion to the ratio of the storage in i at the end of the previous month to the end-of-month storage. Rule B discounts entirely capacities, current projected inflows, and further projected inflows.

Still another rule for the monthly release allocation is

$$(C) \qquad x_{it} = \left\{ \frac{(s_{it-1} + ip_{it})}{\left[\sum\limits_{i=1}^{3} (s_{it-1} + ip_{it}) \right]} \right\} * T$$

Rule C makes the allocations a function of both the storage and the projected inflow for month t, discounting projections through the remainder of the drawdown-refill cycle as well as capacities. One could reinterpret the ip_{it} to be such longer-term projections.

The last rule calculates monthly contributions using storages, projected current inflows, and capacities, but the capacities provide measures of relative as opposed to absolute fullness; that is

$$(D) \qquad x_{it} = \left[\frac{[(s_{it-1} + ip_{it})/c_i]}{\left[\sum\limits_{i=1}^{3} (s_{it-1} + ip_{it})/c_i \right]} \right] * T$$

That is, the proportion of system yield to be taken from reservoir i in month t is given by the portion of system fullness that resides in reservoir i at the end of the month if no release were taken from the reservoir.

As with the space rule, rules (B), (C), and (D) do not provide the system yield T. This number must come from an independent set of calculations. The calculation of system yield T proceeds by operating each reservoir in the system through time using one of the allocation rules (A), (B), (C), or (D) along with mass balance equations. Let us say that rule (A) is selected and the largest system yield is to be determined. Rule (A) is then used in conjunction with the following mass balance equations:

$$s_{it} = s_{it-1} + i_{it} - x_{it} \qquad 0 \le s_{it-1} + i_{it} - x_{it} \le c_i \qquad (2\text{-}10)$$

$$s_{it} = c_i \qquad c_i < s_{it-1} + i_{it} - x_{it} \tag{2-11}$$

$$s_{it} = [\text{inadmissible/stop}] \qquad s_{it-1} + i_{it} - x_{it} < 0 \tag{2-12}$$

These equations are written for each reservoir i in the system for each month of a long record that includes the worst sequence of system flows.

If at any time, using the capacities, inflows, allocation rules, and a tentative value of the system yield — if at anytime, one of the reservoir storages is an inadmissible value, that is, negative, then the system yield T is not attainable with the specified allocation rule.

The means to determine the maximum system yield is to start with a conservative estimate of T, one that is likely to be achievable with the operating rule at hand. A logical number to start with is a value of T that is slightly larger than the sum of the separate historic yields of the individual reservoirs. This number, of course, is likely to be feasible. The system is then operated through all the months of record using this value. Assuming this value of T was feasible, the value of T is increased incrementally until a value of T is reached at which system operation, according to (2-10) to (2-12), leads to a negative storage in some reservoir at some time. When this occurs, the largest T that does not lead to a negative storage anywhere in the system over the entire record is the maximum system yield. A related but different analysis using successive simulation runs for distributing storage and emptiness in reservoirs has been provided by Nalbantis and Koutsoyiannis (1997). They also focus on developing the largest target release for the system. The method of analysis shown in this chapter was first presented by ReVelle (1997, Chapter 1). Another analysis of heuristic operating rules for multiple resevoirs is offered by Johnson and colleagues (1991).

The reader can imagine and create other plausible and useful operating rules than those suggested here, some of which might, under some records, provide a larger system yield than any of the rules described thus far. The apparent quality of one rule vis à vis another could be both record- and capacity-dependent, however.

Optimizing Real-Time Operation and Yield Maximization

MINIMIZING PROJECTED SPILL. In the previous model, we proposed a set of heuristic rules to allocate individual reservoir contributions toward a system target. For a particular rule, we increased the target incrementally until one of the reservoir storages went negative. This was the maximal system yield achievable when the reservoirs were operated according to the heuristic rule. In this model, we once again decide on individual reservoir contributions toward the target and increase the target until some reservoir reaches a negative storage. However, with this model, operation follows a set of optimized monthly decisions rather than any heuristic rule. The key difference here between this and the previous model is that the method for deciding each month on reservoir contributions for that month will be by an optimization model, instead of a rule that follows some intuitive notions of good practice.

The optimization model, which we will assume to use monthly time steps, requires the following notation. We let

s_i = storage in reservoir i at the beginning of the month in which operation is to be taken (also, the storage at the end of the previous month);

x_i = contribution of reservoir i toward the system target;

ip_i = projected inflow in the month to reservoir i

$w_i^{(+)}$ = spill (water wasted) during the month from reservoir i

$w_i^{(-)}$ = expected empty space in reservoir i at the end of the month of operation

c_i = capacity of reservoir i

The projected inflow might be established by use of a regression model that, using historic flows for each reservoir, regresses the flow in month i (say, September) with the flow in month $(i - 1)$ (say, August). Then the projected or forecast value of the flow in month i is calculated by the regression equation using the flow that actually occurred in month $(i - 1)$.

The objective of the optimization model is to deliver the target yield in such a way that when the month is over, the total volume of inflow that is retained in all of the reservoirs is as large as possible. Put another way, maximizing the volume retained is equivalent to minimizing the volume spilled or wasted and places the reservoir system in a strong position to meet future demands. Note that we say "strong position" as opposed to "the strongest possible position" even though we are optimizing for the month. This is because the objective of maximum water retention is a short-term goal. If we are to optimize with a view to the long term, we need a modified objective or a new constraint or set of constraints. We will introduce such a modification once we have structured the basic model. The present model, the one that optimizes monthly operation with an eye to the future, will then be used in conjunction with multiple-system simulations; see Equations (2-10) to (2-12) to establish a maximum system yield.

The reader should take note that the model that allocates reservoir contributions in order to maximize projected water retained/minimize water wasted —with an eye to future needs—is a fundamental model in its own right. It is not simply a model to use in the estimation of maximum system yield. Once a system target has been established for the reservoirs and the reservoirs need to be operated through real, present time, the allocation model we discuss here can be used as the means to make decisions about individual contributions in each month, as that month occurs. That is, the allocation model that we are using here to help us establish the maximum system yield falls also in the class of *real-time decision models.*

The allocation model which minimizes spill may be written as

Maximize $\qquad z = w_1^{(+)} + w_2^{(+)} + w_3^{(+)}$

s.t. $\quad s_i + ip_i - x_i - c_i = w_i^{(+)} - w_i^{(-)} \qquad i = 1, 2, 3 \qquad$ (2-13)

$\qquad\qquad x_1 + x_2 + x_3 = T \qquad\qquad\qquad\qquad$ (2-14)

$\qquad\qquad x_i, w_i^{(+)}, w_i^{(-)} \geq 0 \qquad i = 1, 2, 3$

In the standard form, the model is

Minimize $\qquad z = w_1^{(+)} + w_2^{(+)} + w_3^{(+)}$

s.t. $\quad w_i^{(+)} - w_i^{(-)} + x_i = s_i + ip_i - c_i \qquad i = 1, 2, 3 \qquad$ (2-15)

$\qquad\qquad x_1 + x_2 + x_3 = T \qquad\qquad\qquad\qquad$ (2-16)

$\qquad\qquad x_i, w_i^{(+)}, w_i^{(-)} \geq 0 \qquad i = 1, 2, 3$

By example, we can see that there is often an obvious solution to the problem of minimizing spill. For instance, suppose the three equations (2-15) come out to be

$$w_1^{(+)} - w_1^{(-)} + x_1 = 5$$
$$w_2^{(+)} - w_2^{(-)} + x_2 = -2$$
$$w_3^{(+)} - w_3^{(-)} + x_3 = 5$$

and that

$$x_1 + x_2 + x_3 = 10$$

There are multiple obvious solutions to the problem of minimizing $(w_1^{(+)} + w_2^{(+)} + w_3^{(+)})$. They include the following (x_1, x_3) pairs: (5, 5), (4, 6) (3, 7), with $x_2 = 0$. The methodology is described in Hirsch et al. (1977). All three of the solutions shown can provide zero spill; that is, $w_1^{(+)} + w_2^{(+)} + w_3^{(+)}$.

The possibility of multiple alternate optima always raises the possibility of imposing other objectives or other constraints. In this case, the constraints we will add are designed to select from among the multiple optima those that distribute emptiness in an advantageous way. The allocation-of-emptiness rule is that suggested by Clark, namely, the space rule described by Maas and colleagues (1962) for the distribution of free volume across parallel reservoirs that serve a common supply. Once again, the space rule suggests that the fraction of total system emptiness that resides in an individual reservoir should reflect the fraction of the total system inflow that will occur to that reservoir in some future period of time as measured from the present. Of course, future inflows are predicted inflows, again likely from regression equations.

To achieve Clark's space rule within the context of minimizing the sum of spills, we need to define the fraction of future inflow to reservoir i. This requires a statement of the horizon of concern. If the reservoir refills each year, the horizon may be taken as the number of months prior to all reservoirs in the system refilling. We define

$$\alpha_i = f_i/(f_1 + f_2 + f_3) = \text{fraction of future inflow to reservoir } i$$

where f_i = the anticipated inflow into reservoir i over the planning horizon. Then the constraints that distribute emptiness are

$$\frac{w_i^{(-)}}{\sum_{k=1}^{3} w_k^{(-)}} = \alpha_i \qquad i = 1, 2, 3 \tag{2-17}$$

The mathematical program that minimizes the projected system spill for the month and distributes emptiness at the end of the month is, in standard form,

$$\text{Minimum} \quad z = w_1^{(+)} + w_2^{(+)} + w_3^{(+)}$$

$$\text{s.t.} \quad w_i^{(+)} - w_i^{(-)} + x_i = s_i + ip_i - c_i \qquad i = 1, 2, 3 \tag{2-18}$$

$$x_1 + x_2 + x_3 = T \tag{2-19}$$

$$w_i^{(-)} - \alpha_i(w_1^{(-)} + w_2^{(-)} + w_3^{(-)}) = 0 \qquad i = 1, 2, 3 \tag{2-20}$$

$$x_i, w_i^{(+)}, w_i^{(-)} \geq 0 \qquad i = 1, 2, 3$$

The solution of this linear program provides the release contributions for each of the three reservoirs, contributions that minimize projected system spill in the month and distribute emptiness according to the space rule.

Now that a new mechanism has been carved out for real-time operation, we can return to the maximization of system yield. Once again, we will rely on Equations (2-10) to (2-12) for operation of the reservoir through the historical record. Decisions on individual reservoir contributions for each month are made by the model that minimizes system spill subject to (2-18) to (2-20). Then the actual end-of-period storages are calculated using the historic flows that did occur using (2-10) to (2-12). This operation is performed for a specific value of system yield T. Presumably, the first estimate of T has been conservative, so that no reservoir ever goes negative in storage when real-time operation is practiced and actual historic flows occur. Then T is incremented just until some reservoir storage goes negative over the record of inflows.

Recursive Programming. Our approach to yield maximization has been to specify a system yield and then to determine the monthly contributions toward that yield that are most efficient in some manner, for example in minimizing spill in the upcoming month. Given a system yield, the reservoirs are operated through a long record until such time as the record is exhausted or until some month is reached when the storage in some reservoir becomes negative, based on the allocated contribution. If the record is exhausted without a reservoir storage coming within some tolerance of zero, the yield can be increased and the simulation repeated—using the same procedure for calculating monthly contributions.

So far, we have offered one of three programming/optimization methods for determining the individual reservoir contributions—a mathematical program that minimizes projected spill. The second method, which we now discuss, is recursive or adaptive programming, a methodology specifically designed for operation through time, where the near future is more clearly viewed than the more distant future. In recursive programming, the entire future or an upcoming cycle of events is projected and the values of decision variables through the length of that future or that cycle (say, a year) are determined by a mathematical program. The first month's set of decisions is then conceptually taken or executed. That is, the allocated releases are drawn from the individual reservoirs. At the same time, natural events of the near future, for example, the flows in the first month, then occur (i.e., are read from the long record). The status of the reservoirs is determined via the mass balance equations discussed earlier and the allocation process begins again. By this is meant that the remaining future or the cycle of flows is once again projected and optimal values of decision variables are calculated. The reservoir storages move through time in this way to the end of the long record.

When reservoir storages are updated at the end of each month, each reservoir's contents has been altered by subtracting its contribution toward water supply and subtracting any necessary spill and adding natural inflow. If the system yield is chosen at a feasible level, no storage for any of the reservoirs will ever be calculated as negative at the end of any month. If the system yield has been chosen at an infeasible level, the signal of infeasibility is the occurrence of the first end-of-month negative storage in any reservoir. The use of this approach in calculating maximal system yield is fairly obvious.

Several recursive programming models are possible, depending on the time intervals selected. Our choice for presentation is a model in which at each moment in time (i.e., at the beginning of each month in the long record), 12 successive monthly flows are projected for each reservoir. The method of projection could be regression or could involve snow pack if such a statistic is available. Thus, there are will be multiple "sets of books." There is the *real* book that consists of the long historical record of n monthly flows into each parallel reservoir, a flow history that begins and ends at the same month for each reservoir. Then there are three sets of 12 successive projected monthly flows, one set for each reservoir; these three sets of projected flows begin at the end of each month in the record. How these 12 projected flows are established will determine the first month at which 12 projected flows are available. If for projected flows that begin with month $(t + 1)$, the projection of the flow in the twelfth month ahead in the record requires use of the actual recorded flows in the $(t - 12)$th month, then the first month when 12 projected flows are available will be month 13. In this case, the math program that calculates contributions is first solved for months 13 to 24.

For notation to set up the mathematical program, we will reinterpret the previous notation as follows:

t = index of months in the cycle; $t = 1, 2, \ldots, 12$
i_{it} = projected flow into reservoir i in month t
x_{it} = contribution of reservoir i in month t, unknown
s_{io} = storage in reservoir i at the end of the month that precedes the 12-month cycle, known
s_{it} = storage in reservoir i at the end of month t, unknown
w_{it} = spill from reservoir i in month t, unknown
c_i = capacity of reservoir i, known
T = system yield (or T_t if a monthly system requirement), given

The mathematical program in nonstandard form is

$$\text{Maximize} \quad z = s_{1,12} + s_{2,12} + s_{3,12} \tag{2-21}$$

$$\text{s.t.} \quad s_{it} = s_{it-1} + i_{it} - x_{it} - w_{it} \quad \begin{array}{l} i = 1, 2, \ldots, 3; \\ t = 1, 2, \ldots, 12 \end{array} \tag{2-22}$$

$$s_{it} \leq c_i \quad i = 1, 2, \ldots, 3; t = 1, 2, \ldots, 12 \tag{2-23}$$

$$x_{1t} + x_{2t} + x_{3t} = T \quad t = 1, 2, \ldots, 12 \tag{2-24}$$

$$w_{it}, x_{it}, s_{it} \geq 0 \quad i = 1, 2, \ldots, 3; t = 1, 2, \ldots, 12$$

The constraints correspond to those of the basic reservoir model, but applied to multiple reservoirs. The objective, the maximization of total system storage at the end of the 12th month of the cycle, is designed to place the reservoir system in a very strong position for the remainder of the long record. The objective is approximately equivalent to minimizing spill from the reservoirs over the 12-month cycle.

Once again, the information from this mathematical program is used in the following way in a computer model. A system target T is assumed to be set. Operation proceeds month by month beginning with the thirteenth month of the historical record in a fashion that we illustrate by operation in month t. For that month t in the historical record, there are opening storages, the storages at the end of month $(t - 1)$, that were determined by releases from the reservoirs and actual inflows to the reservoirs during $(t - 1)$.

By using these opening storages and projections of monthly inflows to each of the reservoirs for the next 12 months, the monthly contributions are determined for each of the reservoirs toward the temporarily established system yield. Twelve months worth of monthly contributions are calculated for each of the parallel reservoirs, but the release decisions for the first of the

12 months are the only decisions to be executed. All the rest of the releases are ignored because a new flow now occurs—not the projected flow for the month, but the actual flow as recorded from the historical sequence of flows. This set of flows to each of the reservoirs is treated in combination with the calculated releases from each of the reservoirs to produce a set of storages for the end of the tth month. If any storage is negative, the system yield has been set too high and must be reduced to a feasible level with the entire process repeated from the beginning, that is, from the thirteenth month, to test the new level of system of yield.

Although we have discussed recursive programming in the context of establishing a value for system yield, the technique described is also, as the previous model is, a methodology for real-time operation decisions and could be so used.

Minimizing Expected Spill. The third method of determining the contributions of the individual reservoirs to the system yield and thus affording a way to test the feasibility of a particular value of system yield is a frankly probabilistic optimization model. This model minimizes the expected value of the sum of spills each month where the density function of the monthly inflow to each reservoir is known. We can use analytical density functions and find the minimum by using Lagrange multipliers and differentiating under the integral sign, or we can use approximated density functions and a linear program. The linear programming model to determine the individual reservoir contributions that minimize the sum of spills will be presented here. The reader interested in the method that involves differentiating under the integral is referred to Hadley.[1]

In this model, we are looking only at the upcoming month. Individual reservoir contributions are to be determined for the month in such a way that the expected spill from all three reservoirs is a minimum. Our notation for this model is slightly different than that offered in earlier models.

Let

i_i^k = kth possible realization of inflow to reservoir i during the month, known

p_i^k = probability of the kth possible realization of inflow to reservoir i, known

s_{io} = storage in reservoir i at the end of the previous month, known

s_{ie} = storage in reservoir i at the end of the current month, a random variable

We note that $p_i{}^k$ corresponds to the height of the kth bar in a histogram that approximates the density function for flow into reservoir i during the month. Of course, it must be true that

---

[1] See Hadley, G., *Non-Linear and Dynamic Programming,* Addison-Wesley, Reading, 1964, 484 pages.

$$\sum_{k=1}^{m} p_i^k = 1 \qquad i = 1, 2, \ldots, 3$$

where m is the number of possible realizations used to approximate the density function. Furthermore, i_i^k corresponds to the midpoint of the kth bar in the histogram.

We also define

$U_i^k =$ spill from reservoir i if the kth realization of inflow to i occurs, a random variable

$V_i^k =$ free volume in reservoir i if the kth realization of inflow to i occurs, a random variable

The mass balance relationship can now be written for each reservoir for each possible flow realization. For the ith reservoir and kth realization, the following quantity is defined:

$$s_{io} + i_i^k - x_i - c_i$$

This quantity is positive and represents spill if the beginning storage plus the kth realization of inflow less the contribution of reservoir i toward system yield exceeds the capacity. The quantity is negative and equal to the negative of the free volume if the beginning storage plus the kth realization of inflow less the contribution toward system yield is less than the capacity. Thus, for reservoir i, realization k, the relationship can be written

$$s_{io} + i_i^k - x_i - c_i = U_i^k - V_i^k$$

For each reservoir i, there are m such realization equations.

The mathematical program to minimize expected system spill for the upcoming month is

Minimize
$$z = \sum_{i=1}^{3} \sum_{k=1}^{m} p_i^k U_i^k \qquad (2\text{-}25)$$

s.t. $\quad s_{io} + i_i^k - x_i - c_i = U_i^k - V_i^k \qquad i = 1, 2, 3;\ k = 1, 2, \ldots, m \quad (2\text{-}26)$

$$x_1 + x_2 + x_3 = T \qquad (2\text{-}27)$$

As an alternative objective, one could minimize the expected value of free volume in the system—which maximizes the expected value of the end-of-month system storage. That is,

$$\text{Minimize} \quad z = \sum_{i=1}^{3} \sum_{k=1}^{m} p_i^k V_i^k \quad (2\text{-}28)$$

The solution to either of these mathematical programs provides the monthly contribution of each reservoir toward the prespecified system yield T.

Once again, we have derived a procedure that can be used in conjunction with simulation to establish the maximal feasible value of system yield. This methodology, however, is probably the most labor-intensive of the three methodologies described. As before, a negative value of any reservoir storage when the system is operated through a long record indicates that system yield has been set at an infeasibily high level and needs to be reduced.

Although the minimization of expected system spill can be used to establish the maximum feasible system yield, the procedure, like the two that preceded it, also can be applied to real-time operation of a system of parallel reservoirs. That is, when the system requirement has already been established, the mathematical program provides the allocation formula for release from each of the parallel reservoirs for the month at hand.

COST MINIMIZATION GIVEN A SYSTEM YIELD REQUIREMENT

We indicated earlier that we were first treating the simpler problem, that is yield maximization given reservoirs in place, and that we would return to the obverse or inverted problem. We proceed now to the problem of minimizing cost to obtain a required yield. Because this is already a difficult problem, our treatment will not be as extensive as for the yield-maximization problem. In particular, we will need to make some early assumptions about reservoir operations. There are two possibilities we will consider. The first is that the reservoirs are operated entirely independent of one another. The second is that they will operate in concert with another. This latter assumption is the same as utilized in model (2-1) to (2-8), but leads potentially to a more difficult model to solve. The first leads to a dynamic program of a standard sort.

Reservoirs Operated Independently of One Another

The problem we are stating here is one in which reservoirs are not in place but are to be built, if economically efficient to do so, to provide a specified level of system water supply. As in the earlier system models, we will assume three parallel streams and an associated possible reservoir on each of these streams (see Wathne et al., 1975).

We define

q_T = needed monthly supply of the system, known
q_1, q_2, q_3 = unknown but consistent—through-time monthly supplies to be provided by reservoirs 1, 2, and 3

i_{1t}, i_{2t}, i_{3t} = long record of streamflows by month t for each of the three reservoirs

c_1, c_2, c_3 = unknown capacities of each of the three reservoirs

s_{1t}, s_{2t}, s_{3t},   = unknown storage at the end of month t in each of the three reservoirs

w_{1t}, w_{2t}, w_{3t},   = spills in month t for each of the three reservoirs

$g_i(c_i)$ = cost of reservoir i as a function of its capacity

For each reservoir i, we solve for the smallest capacity subject to Equations (1-1), (1-2), and (1-3), where q_i, the constant monthly release from reservoir i, is initially given; that is,

Minimize $z = c_i$

s.t. $s_{it} = s_{it-1} + i_{it} - q_i - w_{it}$ $t = 1, 2, \ldots, n$ (1-1)

           $s_{it} \leq c_i$ $t = 1, 2, \ldots, n$ (1-2)

           $s_{in} \leq s_{io}$ (1-3)

This is, of course, the fundamental model with which we began Chapter 1. For each reservoir i, we create a storage-yield curve, solving for the needed capacity at increasing levels of reservoir yield q_i. The curve of capacity (as the ordinate) versus reservoir yield (as the abscissa) is not smooth, as we indicated earlier, but is composed of successively higher-sloped linear segments. By using the cost function $g_i(c_i)$ given before, the capacity-versus-yield curve is translated to a cost-versus-yield curve. The new function for reservoir i represented by the curve is $f_i(q_i)$. That is, $f_i(q_i)$ is the cost of the reservoir i needed to deliver a steady monthly supply of q_i.

Now if q_T is demanded from the system each month and each reservoir is to contribute the same quantity each month of the year toward the joint supply, then the problem is

Minimize $z = f_1(q_1) + f_2(q_2) + f_3(q_3)$ (2-29)

s.t. $q_1 + q_2 + q_3 = q_T$ (2-30)

This, of course, is in the standard form of the dynamic program (see, e.g., Dreyfus and Law, 1977). Although the problem stated here is of interest, the assumption of independent operation does not allow for the synergy of joint operation.

Reservoirs Operated in Concert

We choose next to study the problem with the once-and-for-all or zero order rules that were used in model (2-1) to (2-8). These rules allow the contribution of each reservoir toward the common supply to vary by month even though they are the same for a particular month from year to year. In some months, one or two reservoirs may make the dominant contributions, whereas in others, the remaining reservoir(s) may provide the larger amounts toward the system requirement. In this model, as in the preceding model, which used dynamic programming, we avoid entirely the simulation step we had been utilizing in the earlier models. Our notation then will be precisely the notation used in the model (2-1) to (2-8) except that capacities c_1, c_2, and c_3 will be unknown. We will use, in addition, the cost functions $g_i(c_i)$, which indicate reservoir cost as a function of capacity.

The cost of a reservoir increases as the height of the dam is increased. Initially, there is a large "opening cost" and then economies of scale may predominate. Eventually, at larger capacities, the cost curve could become convex. For purposes of this exercise, we approximate the cost-versus-capacity curve as a fixed-charge cost function (see Figure 2-1).

That is, $g_i(c_i)$ is replaced by $b_i y_i + a_i c_i$. The fixed-charge cost function (which will be minimized) states that

$$\text{Cost} = b_i\, y_i + a_i\, c_i \qquad c > 0$$

$$\text{Cost} = 0 \qquad c_i = 0$$

where

y_i 0, 1; it is 1 if reservoir i is built to any level and zero otherwise, unknown

The 0, 1 values of the y_i are achieved in most computer codes by the widely known procedure of Branch and Bound. The b_i and a_i are known constants that define each of the fixed-charge cost functions. Thus, if reservoir i is built to any capacity level other than zero, the cost is $b_i + a_i c_i$. However, if the

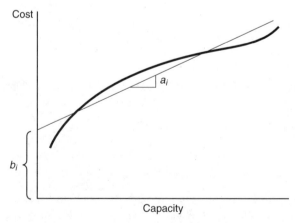

FIGURE 2-1   Fixed-charge approximation of the reservoir cost function.

reservoir is not built, no cost is incurred. In making decisions on what capacities to build to, it is entirely possible that a particular reservoir will not be built at all, and this circumstance must be accounted for correctly by the cost function.

The fixed-charge cost function needs to be enforced by some programming mechanism or constraint. The most common way to enforce this unusual cost function is by introducing an upper bound on the capacity of each reservoir. This upper bound is the largest conceivable capacity that the reservoir could take. We define

m_i = maximum possible capacity for the ith reservoir

The constraint that enforces the fixed charge for the ith reservoir is

$$c_i \leq m_i y_i \qquad i = 1, 2, 3 \qquad (2\text{-}31)$$

This constraint may also be written in a slightly more understandable form as

$$y_i \geq c_i/m_i \qquad i = 1, 2, 3$$

The constraint in combination with a minimizing objective forces y_i to be positive if c_i is positive and zero if c_i is zero. If y_i is forced to be positive by a positive c_i in constraint (2-31), y_i will have to be one since it can only assume the two possible values, zero or one. If c_i is zero, then y_i need only be greater than or equal to zero, and y_i will be pushed down to zero because the objective minimizes the $b_i y_i$ term.

To proceed to minimizing the sum of reservoir costs subject to a monthly system requirement, given once-and-for-all monthly drafts from each of the reservoirs, which drafts may differ by month, we structure the following "mixed-integer" program. By a mixed-integer program, we mean that some of the variables are integer (in this case, zero–one) and some are continuous. Thus, the problem, in nonstandard form, is

Minimize $\qquad z = \displaystyle\sum_{i=1}^{3} (b_i y_i + a_i c_i)$

s.t. $\qquad s_{it} = s_{it-1} + i_{it} - w_{it} - q_{ik} \qquad i = 1, 2, 3;$
$$\qquad\qquad\qquad\qquad t = 1, 2, \ldots, n; k = t - 12[(t - 1)/12] \quad (2\text{-}32)$$

$$s_{it} \leq c_i \qquad i = 1, 2, 3; t = 1, 2, \ldots, n \qquad\qquad (2\text{-}33)$$

$$s_{in} \geq s_{io} \qquad i = 1, 2, 3 \qquad\qquad\qquad\qquad (2\text{-}34)$$

$$q_{1k} + q_{2k} + q_{3k} = q_T \qquad k = 1, 2, \ldots, 12 \qquad\qquad (2\text{-}35)$$

$$c_i \leq m_i y_i \qquad i = 1, 2, 3 \qquad\qquad\qquad\qquad (2\text{-}36)$$

As in model (2-1) to (2-7) and (2-9), if the system requirement were variable by month, q_T would be replaced by Q_A, the known annual requirement, multiplied by β_k, the portion of annual yield needed in month k. That is,

$$q_{1k} + q_{2k} + q_{3k} = \beta_k Q_A \qquad k = 1, 2, \ldots, 12 \qquad (2\text{-}37)$$

and the minimization of costs is carried out subject to all of the earlier constraints except (2-35), which is replaced by (2-37).

Although it is relatively easy to invoke the mixed-integer programming option of most optimization codes, a problem as small as this can be optimized by a combination of linear programming and enumeration. With three reservoirs, there are only seven system possibilities of open/not-open reservoirs. These are triplets of the y_i, namely, (y_1, y_2, y_3). The possible combinations are $(1, 0, 0)$, $(0, 1, 0)$, $(0, 0, 1)$, $(1, 1, 0)$, $(1, 0, 1)$, $(0, 1, 1)$, and $(1, 1, 1)$.

That is, one at a time, some one of the these combinations is examined and the linear portion, as opposed to the integer portion, of the objective is minimized. For instance, when $(1, 0, 1)$ is examined, the objective is $a_1 c_1 + a_3 c_3$ and the constants b_1 and b_3 are added to the value of the optimum linear objective. The least costly of these seven combinations is the answer to the minimization of the fixed-charge objective function. Of course, systems with many reservoirs become more difficult to examine in this enumerative fashion, and the mixed-integer programming option becomes the methodology of choice. With this cost-minimizing model, we conclude our consideration of water supply only models.

CONCLUSION

The justification for spending so much text on water supply only models is that water supply is probably the reservoir use that is first in importance worldwide. Water for municipal and industrial use as well as for agricultural irrigation are among the chief reasons that reservoirs are built. Indeed, there is a great deal for which we have said little about water supply.

We have been cursory in Chapter 1 in our treatment of how much or how frequently water supply releases would be cut back in the event of a shortfall, having said nothing about this topic in this chapter. The study of the reliability of water supplies is ongoing. The consideration of the stochastic or random nature of stream flow began with the work of Fiering and colleagues (Fiering, 1967; Fiering and Jackson, 1971) referred to earlier. Fiering's ideas and time-series notions, in general, were applied by Hirsch (1979) in an examination of reservoir reliability in real time. In addition, the behaviors or responses of reservoirs in the face of random flows, their reliability, resilience, and vulnerability have been examined by a number of investigators including Hashimoto et al. (1982), Moy et al. (1986), and most recently Vogel and Bolognese (1995). Recently, Shih and ReVelle (1994, 1995) began an investigation for the single resevoir to identify the signals to be used for beginning various phases of water rationing. And we have only lightly touched on the issue of reservoir operating rules (see Yeh, 1985). Nor have we been able to say anything about water pricing (see, e.g., Howe and Linaweaver, 1967). And, finally, we have not offered any modifications to incorporate seepage loss. Frankly, in these two introductory chapters on water supply systems, we are unable to elaborate on some truly interesting and valuable research and modeling in water supply planning. This is because there is much still to say about the other functions of water supply reservoirs in addition to water supply and flood control. What we can say, however, with certainty is that the stream of problems to be investigated in water supply planning shows no sign of diminishing over time.

BIBLIOGRAPHY

Dreyfus, S., and A. Law. 1977. *The Art and Theory of Dynamic Programming.* Academic Press, Orlando, Fla.

Fiering, M. 1967. *Synthetic Hydrology.* Harvard University Press, Cambridge, Mass.

Fiering, M., and B. Jackson. 1971. *Synthetic Streamflows,* Water Resources Monograph 1. American Geophysical Union, Washington, D.C.

Hashimoto, T., J. Stedinger, and D. Loucks. 1982. "Reliability, Resilience and Vulnerability Criteria for Water Resource System Performance Evaluation." *Water Resources Research,* Vol. 18, No. 1, pp. 14–20.

Hirsch, R. 1979. "Synthetic Hydrology and Water Supply Reliability." *Water Resources Research,* Vol. 15, No. 6, pp. 1603–1615.

Hirsch R., J. Cohon, and C. ReVelle. 1977. "Gains from Joint Operation of Multiple Reservoir Systems." *Water Resources Research,* Vol. 13, No. 2, pp. 239–245.

Howe, C., and P. Linaweaver. 1967. "The Impact of Price on Residential Water Demands and Its Relation to System Design and Price Structure." *Water Resources Research,* Vol. 3, No. 1, pp. 12–31.

Johnson, S., J. Stedinger, and K. Taschurs. 1991. "Heutistic Operating Policies for Reservoir System Simulation." *Water Resources Research,* Vol 27, No. 5, pp. 673–685.

Maas, A., M. Hufschmidt, R. Dorfman, H. Thomas, S. Marglin, and G. Fair, *The Design of Water-Resource Systems.* Harvard University Press, Cambridge, Mass.

Major, D., and R. Lenton. 1979. *Applied Water Resource Systems Planning,* Prentice Hall, Englewood Cliffs, N.J.

Moy W.-S., J. Cohon, and C. ReVelle. 1986. "A Programming Model for Analysis of the Reliability, Resilience, and Vulnerability of a Water Supply Reservoir," *Water Resources Research,* Vol. 22, No. 4, pp. 489–498.

Nalbantis, I., and D. Koutsoyiannis, 1997. "A Parametric Rule for Planning and Management of Multiple Resevoir Systems." *Water Resources Research,* Vol. 33, No. 9, pp. 2165–2177.

Palmer, R., J. Smith, J. Cohon, and C. ReVelle. 1982. "Reservoir Management in the Potomac River Basin." *Journal of Water Resources Planning and Management (ASCE)*, Vol. 108, No. 1, pp. 47–66.

ReVelle, C., 1997. "Water Resources: Surface Water Systems." In *Design and Operation of Civil and Environmental Engineering Systems,* ed. C. Revelle and A. McGarity. John Wiley, New York.

Shih, J.-S., and C. ReVelle. 1994a. "Water Supply Operations During Drought: A Discrete Hedging Rule." *European Journal of Operational Research,* Vol. 2, pp. 163–175.

Shih, J.-S., and C. ReVelle. 1994b. "Water Supply Operations During Drought: Continuous Hedging Rule." *Journal of Water Resources Planning and Management (ASCE),* Vol. 120, No. 5, pp. 613–629.

Vogel, R., and R. Bolognese, 1995. "Storage–Reliability–Resilience—Yield Relations for Over-Year Water Supply Systems." *Water Resources Research,* Vol. 31, No. 3, pp. 645–654.

Wathne, M., J., Liebman, and C. ReVelle 1975. "Capacity Determination for Water Supply Reservoirs in Series and Parallel." *Water Resources Bulletin,* June.

Yeh, W. 1985. "Reservoir Management and Operations Models: A State-of-the-Art Review." *Water Resources Research,* Vol. 21, No. 12, pp. 1797–1818.

CHAPTER 3

THE HYDROPOWER FUNCTION

The energy in falling water was one of the first forms of energy to be captured and converted to work. The old grist mill has now given way to the modern hydroelectric power plant and the energy that had been used locally may now be transported many hundreds of miles as electricity. Just as at the grist mill though, water turns a "wheel," now called a turbine. The energy that goes into turning the turbine depends on the plant efficiency and on the product of two controllable variables: the volumetric water flow through the turbine and the height of water from the water surface to the turbine. This height is referred to as "head."

ACHIEVING A FIRM ENERGY SUPPLY: SINGLE RESERVOIR

In the case of the water supply reservoir, the question asked was: "How much water can be reliably delivered month after month from a reservoir of stated capacity through a long record that includes droughts?" A perfectly parallel question can be asked for a reservoir devoted to hydropower, namely, "How do we operate a reservoir of a stated capacity to provide the largest amount of energy that can be delivered steadily month after month through a long record that includes droughts?" This is the question we address first in this section. We then expand on this question to focus on the situation in which energy requirements vary in a known fashion month by month, and it is the annual hydropower delivery that is to be optimized. This latter situation parallels the problem we developed for water supply in Chapter 1 in which requirements cycled through the year in a predictable fashion.

Before pursuing the development of the mathematics of hydropower optimization, we need to describe how the power equations are derived. The power equations are really nothing more than the energy equations of basic physics modified by conversion factors that reflect the system of units being utilized in the reservoir model.

If x_t is the release per month, then x_t divided by the seconds in a month is the release per second. The kilowatt-hours of energy produced in a month or season when r cubic meters per second fall through a height in h meters and are released through a turbine with an efficiency factor of ε is

$$2.73 \times 10^{-3}(\varepsilon)(r)(h)\theta$$

where θ is the seconds per month, or season.

If x is the cubic meters released in the month or season, the variable r is x/θ, and the kilowatt-hours of energy produced is

$$2.73 \times 10^{-3}(\varepsilon)(x)(h)$$

where the head is assumed to be maintained through the month or season (see Major and Lenton, 1979, p. 71).

We define

$$\alpha = 2.73 \times 10^{-3}(\varepsilon)$$

and the power produced is now given by

$$\alpha x h$$

In addition to the notation introduced for the basic reservoir model, we need now to add subscripts for time periods to the variables just discussed and we need to define the interval average of head and storage.

h_t = head or height in feet of water above the turbines at the end of month t

$\overline{h}_t$ = average head in month t

$\overline{s}_t$ = average storage volume during month t

x_t = amount of water released through the turbines in month t

The height of water above the turbines is obviously a function of the volume of water stored in the reservoir. The function is represented by the solid curve in Figure 3-1 and is approximated by the dashed line, which may be written as the linear function

$$h = h_o + ms$$

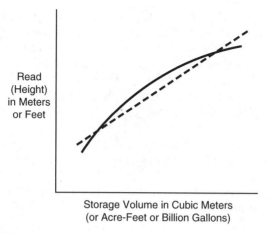

Read
(Height)
in Meters
or Feet

Storage Volume in Cubic Meters
(or Acre-Feet or Billion Gallons)

FIGURE 3-1   Head as a function of storage volume.

where h and s are the instantaneous height and storage, respectively. Thus, the average height in month t may be written as a function of the average storage in month t

$$\bar{h}_t = h_o + m\bar{s}_t \tag{3-1}$$

The average storage in month t may be well approximated by

$$\bar{s}_t = (s_{t-1} + s_t)/2 = (1/2)s_{t-1} + (1/2)s_t \tag{3-2}$$

The hydroelectric energy generated in month t is proportional to the product of average head during t and release throughout t, as discussed earlier.

$$\pi_t = \alpha\bar{h}_t x_t$$

where π_t is the hydroelectric energy generated in month t.
   Substituting Equation (3-1) for $\bar{h}_t$ gives

$$\pi_t = \alpha(h_o + m\bar{s}_t)x_t \tag{3-3}$$

In Equation (3-3), $\bar{s}_t$ can be replaced by its equivalent in Equation (3-2), so that energy production in t can be written as

$$\pi_t = \alpha[h_o + (m/2)s_{t-1} + (m/2)s_t]x_t \tag{3-4}$$

It is this equation for hydroelectric energy production that is used in conjunction with the basic reservoir model (1-1), (1-2), and (1-3).

The question that is being asked first, the maximum amount of firm, steadily deliverable energy that can be provided from a reservoir of capacity c, cannot be approached directly with linear programming, to our knowledge. Nonetheless, by solving a linear program repeatedly, that value, the firm energy, can be determined relatively precisely. It is of interest that, although the calculation of maximum firm energy requires multiple solutions of the linear program, the calculation of the minimum storage needed to deliver a steady month-to-month energy requirement is immediate. Only one linear program need be solved to derive this information. As a consequence, a storage-energy curve (as opposed to storage-yield curve) can be easily calculated. We explain how to do this shortly.

We begin by prespecifying an amount of firm energy delivery at a level we are confident can be achieved:

d = minimal monthly energy production

We then write the basic reservoir constraints and append a set of constraints to them that requires that the prespecified energy be produced in each month. These appended constraints are nonlinear. The model constraints take the form

$$s_t = s_{t-1} + i_t - x_t - w_t \qquad t = 1, 2, \ldots, n \qquad (3\text{-}5)$$

$$s_t \le c \qquad t = 1, 2, \ldots, n \qquad (3\text{-}6)$$

$$s_n \ge s_o \qquad (3\text{-}7)$$

$$\alpha[h_o + (m/2)s_{t-1} + (m/2)s_t]x_t \ge d \qquad t = 1, 2, \ldots, n \qquad (3\text{-}8)$$

The challenge is to find a feasible solution to (3-5) to (3-8) that produces energy at a rate d. The problem has been thought to be difficult because constraints (3-8) involve product terms. This model differs from the basic model (1-1), (1-2), and (1-3), not only because of the presence of the nonlinear energy production constraints (3-8), but also in the release quantity x_t, which is allowed to differ for each month of the entire record. The fact that the release quantity x_t differs throughout the record establishes the firm energy that is determined as a theoretical maximum. In practice, an operator would not be expected to face the same inflow record ever again and so would not be able to follow the optimal sequence of releases prescribed by the mathematical program. After the methodology is described, we show how to determine a practical maximum of firm energy. It should be noted that this release x_t is not for the water supply, but for power production and that the quantity of water that flows through the turbines becomes downstream flow.

To find a feasible solution to (3-5) to (3-8), we first transform (3-8) into linear constraints. Constraints (3-8) can also be written

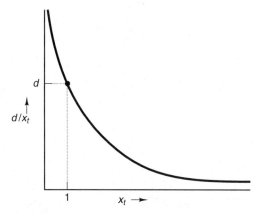

FIGURE 3-2   The equation d/x_t as a function of x_t.

$$\alpha h_o + (\alpha m/2)s_{t-1} + (\alpha m/2)s_t \geq d/x_t \qquad t = 1, 2, \ldots , n \qquad (3\text{-}9)$$

The term (d/x_t) is a nonlinear function whose shape is illustrated in Figure 3-2. It is an hyperbola that approaches zero as x_t becomes large and that approaches infinity as x_t goes to zero.

We note that the function shown in Figure 3-2 is convex because a line segment joining any two points on the curve falls entirely above the curve except for its end points. Constraints (3-9) can be reorganized to read

$$d/x_t - (\alpha m/2)s_{t-1} - (\alpha m/2)s_t \leq \alpha h_o \qquad (3\text{-}10)$$

This constraint has two linear terms and a convex term and is less than or equal to a right-hand side value. The linear terms are standard issue for the application of linear programming. The convex term, because the sense of the constraint is "less than," can be piecewise approximated by line segments and the segments will enter in the proper order. In Figure 3-3, we show the piecewise approximation of (d/x_t) in three segments. The approximation will be the same for all months. In the figure, the following new notation is introduced:

r_o = value of d/x_t at the first point of approximation
r_i = absolute value of the slope of the ith segment
x_{it} = the amount of the ith segment filled or occupied
l_i = the length of segment i

With this new notation, constraints (3-10) are replaced by constraints (3-11), (3-12), and (3-13):

$$r_o - \sum_{i=1}^{3} r_i x_{it} - (\alpha m/2)s_{t-1} - (\alpha m/2)s_t \le \alpha h_o \qquad t = 1, 2, \ldots, n \quad (3\text{-}11)$$

$$x_t - \sum_{i=1}^{3} x_{it} = l_o \qquad t = 1, 2, \ldots, n \qquad (3\text{-}12)$$

$$x_{it} \le l_i \qquad \begin{aligned} &t = 1, 2, \ldots, n; \\ &i = 1, 2, 3 \end{aligned} \qquad (3\text{-}13)$$

in which constraints (3-11) force hydroenergy production in each month to be greater than or equal to d; constraints (3-12) define x_t in terms of the x_{it}; and constraints (3-13) limit the x_{it} to the appropriate length.

The new constraint set consists of (3-5), (3-6), (3-7), (3-11), (3-12), and (3-13). We are assuming here that sufficient turbine capacity is in place that the entire potential monthly energy production can, in fact, be generated. The value of d was chosen at a conservative level, a level we were confident could be delivered. Note that no objective function has been articulated other than the maximum firm power, but the firm power is temporarily a specification rather than the objective. To be certain that the initial value of d can be delivered, we need to solve (3-5) to (3-7) and (3-11) to (3-13) under some objective. Any reasonable objective will do. As an example, we might maximize the average storage, that is,

$$\text{Maximize} \qquad z = \sum_{t=1}^{3} s_t$$

If the solution to this problem is feasible, that is, a energy level of d can be delivered, we will obtain an answer that consists of n values of the x_t, $(n + 1)$ of s_t, and n of w_t.

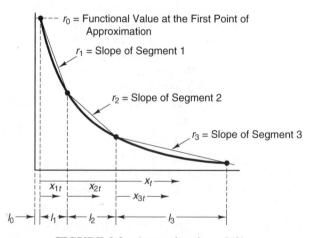

FIGURE 3-3   Approximation of d/x_t.

If it is feasible to deliver a firm energy of d, then the next step is to increase d by a small increment. For this new value of d, the function d/x_t will need to be reapproximated. Probably the segment lengths can remain the same, but the slopes of the approximating line segments will need to be recalculated for each new value of d. The firm energy level is increased in steps until the linear program is no longer feasible. The highest firm energy level that can be produced through the historical record from a reservoir of capacity c is the last value of d attempted before the next increment produces an infeasible solution. For each value of capacity, a maximum firm energy level can be calculated in this manner.

The Storage-Energy Curve

The problem may be also efficiently approached in the following way. Set a firm energy level d. Using piecewise approximation, write the constraints requiring that power level. Minimize the capacity necessary to achieve this. Of course, if d is chosen too high, no capacity will be capable of providing it and the problem will be infeasible. Let's assume though that we have specified a value of d that can be achieved. If that is the case, we can calculate the smallest needed capacity to accompany the value of d. We then specify multiple values of d from small to large. For each value of d, each of which requires reapproximation of d/x_t, a capacity is determined. Now a figure can be created in which d is plotted against c, firm energy against capacity. The plot, a storage-energy curve, is analogous to the storage-yield curve we described in the first portion of the chapter.

The question we asked earlier—what is the largest firm energy that can be delivered from a reservoir of a stated capacity?—can now be answered quite simply by referring to the curve. For the given capacity, the largest firm energy can be read directly from the curve, as could the firm energy level at any other capacity.

Maximizing the firm energy may presuppose that the need for the energy is constant in time. This is analogous to the situation in the water supply models where an historic yield, constant through the year, was determined given a reservoir capacity, or where capacity was minimized with a specified constant monthly yield. We showed how, in the water supply models, to disaggregate an annual requirement into monthly requirements and then how to minimize capacity subject to the annual requirement with its particular seasonal needs. An analogous procedure for hydroenergy is now described in which the energy demand to be met changes with the month.

We define

d_A = annual hydroenergy production, specified

δ_k = portion of the annual hydroenergy production needed in month k of the year given

$\delta_k d_A$ = hydroenergy need in month k

The index of k is from 1 to 12 rather than from 1 to n, and the sum of the δ_k over the 12 months is unity.

The hydroproduction constraint (3-10) is rewritten to reflect the monthly need as follows:

$$(\delta_k d_A)/x_t - (\alpha m/2)s_{t-1} - (\alpha m/2)s_t \leq \alpha h_o \qquad (3\text{-}14)$$

where, for a given t,

$$k = t - 12[(t - 1)/12]$$

$$[u] = \text{integer part of } u \text{ or largest integer inside of } u$$

Once again, the function $\delta_k d_A/x_t$ is an hyperbola and is convex so that piecewise approximation of this term of the constraint (3-14) will, when solved as part of a linear program, yield variables entering in the correct order.

There are now 12 different hyperbolae with which to contend, one for each month, and each must be separately approximated. Thus, the slope of the ith segment is now a function of the month's hyperbola being approximated. However, we can leave the point of first approximation at the same value for each month and the segment lengths the same as well. Thus, we have

r_{ik} = absolute value of the slope in the ith segment for month k

Constraint (3-11) now becomes

$$r_o - \sum_{i=1}^{3} r_{ik}x_{it} - (\alpha m/2)s_{t-1} - (\alpha m/2)s_t \leq \alpha h_o \qquad t = 1, 2, \ldots, n \quad (3\text{-}15)$$

and constraints (3-12) and (3-13) remain the same.

Now when we minimize capacity subject to (3-5), (3-6), (3-7), (3-12), (3-13), and (3-15), we are finding the least capacity needed to deliver d_A total units of electrical energy that are distributed across the 12 months via a set of coefficients δ_k.

As before, we can vary the annual energy yield d_A and, for each value, find the smallest reservoir capacity. These data can be used to create a storage-energy curve as before but with the energy being the annual production rather than a monthly production. Again, however, the energy yield is a theoretical maximum.

Earlier in the chapter, we indicated that this is a theoretical maximum firm hydroenergy because the historical record will never repeat precisely. As a consequence, the operating guidance we derived for either situation, the energy yield constant across the year or the energy yield varying with month, can likely never be directly applied to future flow situations. We need then some form of guidance that can provide the operator signals for how much to release through the turbines each month.

Historically, reservoir operators have guided hydropower production by aiming for a storage level particular to the month or season. Thus, 12 monthly storage levels may be specified for each month of the year, levels at which each month begins and a level at which each month ends. The plot of storage target through time is, of course, the familiar rule curve for reservoir operation. The model for hydropower production will now be modified to produce such a rule curve.

We will describe the methodology for the situation in which an annual energy demand is disaggregated by month because the notational needs of the new model dovetail with the notation introduced in the annual demand model. Once again, this model, which will produce a rule curve, is intended to produce ultimately a practical, as opposed to theoretical, level of annual energy production, that is, a level of annual energy production likely to be achievable if the future resembles but does not duplicate the past. Of course, if it is possible to generate the firm energy in a given month and still have spare storage, then the operator could accumulate additional storage up to the capacity for the next month if it was desirable.

Since a rule curve provides a series of end-of-month storages, the first step of our procedure is to declare that there are only 12 such storages and not $n + 1$. We use s_k to replace s_t in the mass balance equation (3-5) and in the storage capacity equation (3-6) and the can't-borrow-water equation (3-7) and in the hydropower production equation (3-15).

$$s_k k = s_{k-1} + i_t - x_t - w_t \qquad t = 1, 2, \ldots, n;$$
$$k = t - 12[(t - 1)/12] \qquad \qquad (3\text{-}16)$$

$$s_k \leq c \qquad k = 1, 2, \ldots, 12 \qquad \qquad (3\text{-}17)$$

$$s_{12} = s_o \qquad \qquad (3\text{-}18)$$

$$r_o - \sum_{i=1}^{3} r_{ik} x_{it} - (\alpha m/2) s_{k-1} - (\alpha m/2) s_k \leq \alpha h_o \qquad t = 1, 2, \ldots, n;$$
$$k = t - 12[(t - 1)/12] \quad (3\text{-}19)$$

and, of course, add the definitional constraints for the x_{it}, namely,

$$x_t - \sum_{i=1}^{3} x_{it} = l_o \qquad t = 1, 2, \ldots, n \qquad \qquad (3\text{-}12)$$

$$x_{it} \leq l_i \qquad t = 1, 2, \ldots, n \qquad \qquad (3\text{-}13)$$

Constraint (3-18) indicates that the opening storage (end of month is zero) and closing storage (end of month is 12) must be the same.

If reservoir capacity is an unknown, it shifts to the left-hand side of constraint (3-17) and can be minimized. Minimizing capacity for various levels of annual demand (distributed by a set of coefficients across the months) produces the same sort of storage-energy curve as described earlier with one difference. The energy levels are practical, likely to be producible, levels of annual energy. These levels should be nearly capable of production if the storage values are adhered to each month. Of course, a sharp shift in storage values from one month to the next is not what is expected. A roughly linear change in storage level from one month to the next is probably a reasonable strategy to which to aspire.

If a reservoir were already in place and turbines already installed, then the energy production in month t would be limited by the products of hours in the month and the installed capacity (in power units, kW or MW) of the turbines.

Before leaving the question of achieving the minimum capacity reservoir that achieves some firm annual or monthly energy level, we need to reflect on what the releases through the turbines should be when the rule curve is implemented in a real situation. The x_t values from the solution to the linear program are not applicable in the world of real-time inflows. We let

$$s_{t-1}^* = \text{rule curve storage at the end of month } t - 1$$
$$s_t^* \ \ = \text{rule curve storage at the end of month } t$$
$$\theta \ \ = \text{fraction of the month elapsed}$$

At some moment θ within month t, the storage should be

$$s_t(\theta) = s_{t-1}^* + (s_t^* - s_{t-1}^*)\theta \qquad \text{if } s_t^* > s_{t-1}^*$$
$$s_t(\theta) = s_{t-1}^* - (s_{t-1}^* - s_t^*)\theta \qquad \text{if } s_{t-1}^* > s_t^*$$

One would anticipate that, starting from the storage at the end of month $t - 1$, that the release through the turbines and the release as spill that are needed to achieve $s_t(\theta)$ would be calculated by

$$s_t(\theta) = s_{t-1} + i_t(\theta) - x_t(\theta) + e_t(\theta)$$
$$x_t(\theta) + w_t(\theta) = s_t(\theta) - s_{t-1} - i_t(\theta)$$

If $[x_t(\theta) + w_t(\theta)]/\theta$ is greater than the turbine capacity, then the monthly turbine capacity, call it f, governs, and $f\theta$ is the limit on $x_t(\theta)$ so that

$$w_t(\theta) = s_t(\theta) - s_{t-1} - i_t(\theta) - f\theta$$

The problem of maximizing the cumulative hydroenergy production over a year from a single reservoir is not so straightforward as maximizing the firm yield. Many investigators have studied the problem (see the Bibliography at end of chapter). Nor is the problem of optimizing firm electric energy yield from multiple reservoirs, in which the hydroenergy constraint consists of the sum of product terms, is an easy problem. These two topics are the next foci of the chapter.

MAXIMIZING ANNUAL PRODUCTION OR REVENUES: SINGLE RESERVOIR

Annual production of hydroelectric energy or annual revenues from the sale of hydroelectric energy are really the same problem, but in the latter, the monthly production is weighted by its value in the electricity marketplace in that month. Even though most electric energy prices do not change by month (although seasonal changes are common), we can use

v_t = selling price of a kilowatt-hour in month t

because we can allow v_t to be one value in the summer months and another in the winter months. We will only structure the problem of maximizing revenues, knowing that if all values of v_t are equal, we will be maximizing annual hydroenergy production.

The problem can be structured using the previous notation as follows:

$$\text{Maximize} \quad z = \sum_{t=1}^{n} v_t \pi_t = \sum_{t=1}^{n} v_t [\alpha(h_o + (m/2)s_{t-1} + (m/2)s_t)x_t] \quad (3\text{-}20)$$

$$\text{s.t.} \quad s_t = s_{t-1} - x_t - w_t + i_t \quad t = 1, 2, \ldots, n \quad (3\text{-}5)$$

$$s_t \leq c \quad t = 1, 2, \ldots, n \quad (3\text{-}6)$$

$$s_m \geq s_o \quad (3\text{-}7)$$

Note that constraints (3-5), (3-6), and (3-7) are precisely those of the previous model and that the hydropower production constraints (3-11), (3-12), and (3-13) could be added if there was a lower bound on monthly production.

The objective (3-20) when expanded is the sum of products of decision variables plus a term that is linear in only x_t. The method employed in solving the problem of minimizing capacity given a hydroenergy constraint can no longer be used since that method relied on separation of variables by a division step—a step that will not work with the objective (3-20). Thus, we must approach the problem in a different way.

A common approach to nonlinear problems containing inequations with products of variables is to estimate those variables that, when valued, make the inequations linear. At the same time, solution of the linearized problem with its inequations, returns new values of those estimated variables for reentry in the original problem. If and when the returned values match the previously entered estimates within a specified tolerance, the solution process has reached a stable point, a point from which further iterations are unlikely to yield a closer match of entered and returned variables. The stable point is often a "good" solution to the original nonlinear problem. No claim of optimality is likely to be made or should be made; only local optimality or stability can be declared. That is, the constraints are satisfied and the objective is at a locally high or low point. It is possible that another set of decision variables could also satisfy the inequations and provide a better value of the objective.

This is the approach to be discussed here. In the present case, one set of variables is probably most efficient to set and reestimate the storage variables. To see why, we return to the original problem described by (3-20), (3-5), (3-6), and (3-7). Suppose we could solve this problem exactly by some methodology. That is, we are supposing that a globally optimal solution could be found. The output would contain a sequence of storage, release, and spill values. The value of the objective would be maximal. The solution, however, would not be implementable and does not provide much guidance to the operator because the historical record on which it is based is not to be expected to ever occur again. That is, the decision variables would be of little use for future decisions. Thus, the objective is an overestimate of that power that could be practically achieved in the future. As we discussed in our treatment of a firm energy requirement, the approach followed in the world of practical reservoir operation is the creation of a rule curve, a set of end-of-month storages for each month of the year, which sequence of storages repeat each and every year. That is the approach suggested here—the derivation of a storage rule curve.

To produce the new problem, we restate (3-20), (3-5), (3-6), and (3-7) with only 12 storage variables:

$$\text{Maximize} \quad z = \sum_{t=1}^{n} v_k \left[\alpha(h_o + (m/2)s_{k-1} + (m/2)s_k)x_t \right]$$
$$k = t - 12[(t-1)/12] \qquad (3\text{-}21)$$

$$\text{s.t.} \quad s_k = s_{k-1} - x_t - w_t + i_t \qquad t = 1, 2, \ldots, n;$$
$$k = t - 12[(t-1)/12] \qquad (3\text{-}22)$$

$$s_k \leq c \qquad k = 1, 2, \ldots, 12 \qquad (3\text{-}23)$$

$$s_{12} = s_o \qquad (3\text{-}24)$$

The fact that there are only 12 storage variables in this new problem provides a strong incentive to use these variables for the estimation-reestimation process. The process might begin with a specification of 12 equal and relatively low storage values that would be placed in the objective as indicated. The resulting linear program would be solved and 12 new estimates of storage returned since the storage variables are essential parts of the constraint set.

These new estimates would then be placed in the objective, providing a new linear objective. The process would be repeated until

$$s_k^r - s_k^{r-1} < \varepsilon \qquad k = 1, 2, \ldots, 12$$

$$s_k^{r-1} - s_k^r < \varepsilon$$

where s_k^r is the storage at the end of month k on iteration r, and ε is some tolerance level within which changes must occur for the process to terminate. The termination rule is that the difference between all storage variables from one iteration to the next be less than some tolerance. A convergence to a set of stable values for the s_k is the desired outcome, but the author's practical experience suggests that convergence does not always occur.

Instead of the variables converging to stable values, they may "bounce around," moving back and forth across the region in which they are likely to lie. Such behavior is, in fact, not uncommon for such problems and is quite descriptively referred to as zigzagging. The occurrence is so common that procedures have evolved to damp out the zigzagging. They are also colorfully described as anti-zigzagging. The essence of anti-zigzagging procedures is a truncation of the step that seems to be called for by the reestimation process. We describe anti-zigzagging by example.

Suppose we have a process in which the variables are expected to bounce around. We begin with s_k^0, the initial estimate of the kth storage. The temporary calculated value from a run of the linear program that used the initial estimate is $\hat{s}_k^1$ where ^ indicates temporary or tentative and the one indicates that the value comes from run 1. The value to be used in the second linear program, using a typical antizig-zagging formula (there are many), is

$$s_k^1 = s_k^0 + (\hat{s}_k^1 - s_k^0)/2$$

That is, the estimate to be placed in the second linear program is a value that is half the "distance" between the initial estimate and the estimate from the linear program. On the next step, the value to be entered in the linear program is

$$s_k^2 = s_k^1 + (\hat{s}_k^2 - s_k^1)/3$$

We formally define

s_k^r = estimate of s_k entered in the $(r + 1)$st linear program. It is an adjusted value from the rth linear program

$\hat{s}_k^r$ = tentative value of s_k on the rth step (from the rth linear program). It is the value from which the final estimate s_k^r is calculated.

Then we might use the following formula:

$$s_k^r = s_k^{r-1} + (\hat{s}_k^r - s_k^{r-1})/(r + 1) \tag{3-25}$$

so that

$$s_k^1 = s_k^o + (\hat{s}_k^1 - s_k^o)/2$$
$$s_k^2 = s_k^1 + (\hat{s}_k^2 - s_k^1)/3$$

and so on. The zigzagging and anti-zigzagging processes are illustrated in Figure 3-4.

Note that the step size is forced to decline as iterations continue. The rate of step size reduction can be controlled by adjustment of the denominator utilized in the reestimation formula. Anti-zigzagging may require some adjustment of the denominator depending on the characteristics of the particular problem, but, in general, the procedure damps changes down efficiently and leads to stable solutions.

Zigzagging is a potential problem with the estimation-reestimation process, but it is curable as we describe. Another difficulty in this problem occurs because of the product terms in the objective. The product terms lead to a boom-and-bust phenomena in the operation of the reservoir. Many runs, actually sequences of linear programs with reestimated storage variables, terminate with storages that are at capacity at the end of one month and crash to near zero by the end of the next. There follows a sequence of months with low hydroelectric output (depending on constraints) and a gradual buildup of storage to capacity or near capacity. What is happening is that the product of release times height is so large in the month of depletion/emptying that the rest of the year is written off for much power generation.

To correct for this phenomenon, one can add the firm energy constraints (3-11), (3-12), and (3-13) to the formulation. These constraints, by requiring

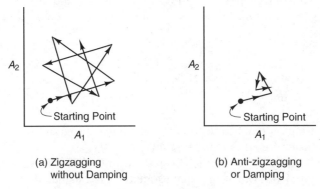

(a) Zigzagging
without Damping

(b) Anti-zigzagging
or Damping

FIGURE 3-4   Zigzagging and anti-zigzagging illustrated by a problem with two variables.

a base level of hydroelectric output, pull down the peak production and flatten the trace of electric energy output.

Once again, if turbines are already in place, energy production for month t is limited by the product of installed capacity and hours in the month.

VARIABLY PRICED ENERGY OUTPUT: SINGLE RESERVOIR

The previous model sought the largest cumulative energy production, or if prices multiply the monthly production, the maximum revenue from the scale of power. The pricing assumption was that price was particular to each month or season and that all the energy produced in a month could be sold at that price. The hydropower may only be used for peak loads, a common manner of use in the United States. If it *is* being used for peak loads, the price in each month or season is the peak price. The electric energy however, may be used for the base load—in which case, firm energy is the important element. Firm electric energy, the electric energy that can be reliably produced each month, may command a better price than excess energy (energy in excess of firm energy) in contracts for the output of the reservoir. Hence, we need a way to value and count firm energy and excess energy.

We can rewrite the hydropower constraint (3-8) as an equality by subtracting a surplus variable from the left-hand side. At the same time, the right-hand side will now be called F, a variable to be calculated, the firm energy production.

$$\alpha[h_o + (m/2)s_{t-1} + (m/2)s_t]x_t - E_t = F$$

where

E_t = excess energy above F produced in month t
F = lowest energy production in any of the months

Now the objective places a value on firm energy at one price and excess (or "dump") energy at another:

p_E = contract price for excess energy
p_F = contract price for firm energy

The objective is

$$\text{Maximize } Z = p_E \sum_{t=1}^{n} E_t + np_F F \qquad (3\text{-}26)$$

and the constraints are

$$\alpha[h_o + (m/2)s_{t-1} + (m/2)s_t]x_t - E_t = F \qquad t = 1, 2, \ldots, n \qquad (3\text{-}27)$$

$$s_t = s_{t-1} - x_t - w_t + i_t \qquad t = 1, 2, \ldots, n \qquad (3\text{-}5)$$

$$s_t \leq c \qquad t = 1, 2, \ldots, n \qquad (3\text{-}6)$$

$$s_n \geq s_o \qquad (3\text{-}7)$$

This problem, however, is not yet in the form that we want to solve. As in the problem described by (3-20), (3-5), (3-6), and (3-7), a globally optimal solution to this problem would not be implementable because the historical record on which it is based will never be seen again. Instead, the reservoir operator will see a new set of flows, flows with similar statistical properties, but different in other ways. Since an implementable solution is the aim, we return to the idea of creating a storage-rule curve for the guidance of the reservoir operator, a set of end-of-month storages for each of the 12 months of the year.

The problem (3-26), (3-27), (3-5), (3-6), and (3-7) can be restated with only 12 storage variables.

$$\text{Maximize} \qquad Z = p_E \sum_{t=1}^{n} E_t + np_F F \qquad (3\text{-}28)$$

$$\alpha[h_o + (m/2)s_{k-1} + (m/2)s_k]x_t - E_t = F \qquad \begin{array}{l} t = 1, 2, \ldots, n; \\ k = t - 12[(t-1)/12] \end{array} \qquad (3\text{-}29)$$

$$s_k = s_{k-1} - x_t - w_t + i_t \quad t = 1, 2, \ldots, n; \; k = t - 12\,[(t-1)/12] \qquad (3\text{-}30)$$

$$s_k \leq c \qquad k = 1, 2, \ldots, 12 \qquad (3\text{-}31)$$

$$s_{12} = s_o \qquad (3\text{-}32)$$

Of course, F is an unknown and shifts to the left side for the solution process.

As in the problem of maximizing cumulative production or total revenue, the $12s_k$ values are estimated for use—this time in Equation (3-29), which becomes linear when the s_k values are inserted. The linear program is solved, returning 12 new values of the s_k, and the new values are placed in (3-29) in preparation for resubmission of the LP. A second LP is solved, new values of the s_k result, and the process continues until the s_k values stabilize. If zig-zagging occurs, that is, if the s_k values do not stabilize, the changes in the s_k can be damped via Equation (3-25) to decrease the zig-zagging and limit the number of steps needed to stabilize the s_k values. The result, when stable s_k are achieved, is a storage-rule curve that maximizes the revenues from the sale of firm and dump energy.

HYDROPOWER PRODUCTION IN MULTIPLE RESERVOIRS

The situation of reservoirs or parallel streams described in Chapter 2 could apply in the production of electric power. Each reservoir on the parallel streams would be making a contribution of electric energy toward some system target. Here, we extend the firm supply concept to multiple reservoirs and suggest a method to approach the problem.

We first assume that the reservoirs are in place with given capacities. In a second model, we relax this assumption. We use much notation from Chapter 2 and define several new variables applicable to multiple reservoirs. As in Chapter 2, we assume for simplicity that there are three parallel reservoirs.

x_{it} = release through the turbines of reservoir i in month t, unknown
c_i = capacity of reservoir i, known
i_{it} = inflow to reservoir i in month t, known
w_{it} = spill from reservoir i in month t, unknown
s_{it} = storage in reservoir i at the end of month t, unknown
h_{it} = head above the turbines, reservoir i, month t, unknown
$\overline{h}_{it}$ = average head during t in reservoir i, unknown
α_i = power factor for reservoir i
h_{oi}, m_i = intercept and slope of the height vs. storage equation for reservoir i

As earlier, the average head is given by the equation

$$\overline{h}_{it} = h_{oi} + m_i(s_{it-1} + s_{it})/2$$

The energy produced in period t is the sum of the energies produced in the individual reservoirs. In equation form, this is the sum of the terms of (3-4), which sum is required to be greater than or equal to the firm energy supply D.

$$\sum_{i=1}^{3} \alpha_i[h_{oi} + (m_i/2)s_{it-1} + (m_i/2)s_{it}]x_{it} \geq D \qquad t = 1, 2, \ldots, n \quad (3\text{-}33)$$

When we modeled the firm energy for the single reservoir, we were able to divide both sides of Equation (3-8) by the flow through the turbine. In Equation (3-33), we are not able to find a single variable by which we can divide. As a consequence, the linearization that we performed on Equation (3-8) is not available to us.

Approaches to this problem are described in Major and Lenton (1979) and in Loucks et al. (1981) but we present two approaches different from these. The first approach is an extension to the methodology we described when we maximized annual hydropower revenue or production, namely, the determination of a storage-rule curve, but in this case, we produce a rule curve for each reservoir. This first approach is appropriate not only when the reservoirs are devoted solely to hydropower, but also when the reservoirs have other functions beside the production of hydropower. Recreation or water supply are examples of other functions or uses for which this approach would be appropriate.

The second approach, in brief, uses dynamic programming in the same way that dynamic programming was recommended for utilization for multiple reservoirs used only for water supply. That is, the individual reservoirs are assumed to have only a single function: water supply in the original treatment and electric power production in the case to be treated here. It should be added that the second approach makes additional assumptions, but these will be detailed shortly.

There is serendipity in the first approach. The step of saying that a storage-rule curve is the desired output not only provides operating guidance for the reservoir manager; it also simplifies the problem by reducing one facet of the problem's dimensionality. The n storage variables for each reservoir are reduced to 12 for each reservoir.

We now rewrite Equation (3-33) with just 12 storage variables for each reservoir and the remainder of the constraints that define the problem.

$$\sum_{i=1}^{3} \alpha[h_{oi} + (m_i/2)s_{ik-1} + (m_i/2)\, s_{ik}]x_{it} \geq D \quad \begin{array}{l} t = 1, 2, \ldots, n; \\ k = t - 12[(t - 1)/12] \end{array} \quad (3\text{-}34)$$

$$s_{ik} = s_{ik-1} - x_{it} - w_{it} + i_{it} \quad \begin{array}{l} t = 1, 2, \ldots, n; \\ k = t - 12\,[(t - 1)/12] \end{array} \quad (3\text{-}35)$$

$$s_{ik} \leq c_i \quad i = 1, 2, 3 \quad k = 1, 2, \ldots, 12 \quad (3\text{-}36)$$

$$s_{i12} - s_{io} = 0 \quad i = 1, 2, 3 \quad (3\text{-}37)$$

Some straightforward objective, such as maximum average storage across all the reservoirs, or perhaps maximum firm energy, D, is superimposed on the problem. Then the s_{ik}, 36 of them, are estimated for Equation (3-34), and the linear program is solved. New estimates of the s_{ik} are returned from solution of the LP, and the process of specification-reestimation is repeated either until the s_{ik} values have stabilized or it is clear that they are not going to stabilize without outside help. In the latter situation, we perform our previously described antizig-zagging maneuvers to force the storage values toward a stable solution.

The reader may wish to extend this multireservoir model to (1) energy demands that vary through the year or (2) the problem of maximizing annual production or revenue from the system of reservoirs. There remains as well the need to model the multireservoir system that maximizes the revenue from the sale of both firm and dump power.

Finally, there is the problem of designing a system "from scratch" to deliver a steady amount of hydroelectric energy through the year. This is similar to a problem we dealt with in Chapter 2. There we structured a problem to minimize the cost of a reservoir system subject to a constraint that a certain monthly (or possibly annual) system water supply requirement was in place. That model forms the basis for the second of the two modeling approaches we will present for multiple-reservoir hydropower optimization.

We indicated that we would provide two approaches to dealing with electric power production from multiple reservoirs. Although the first methodology we described is complex, it does allow the modeler to analyze or constrain other functions and optimize other objectives. It also allows the contribution of each reservoir toward system firm supply to vary with the month of the year. However, although antizig-zagging is an effective convergence method, the stable solution it locates may not be optimal; at least we have no way to show that it is optimal. All we know is that the solution is stable and probably good. Convergence to local optimal/stable solutions is as well a characteristic of the procedures described in Major and Lenton (1979) and Loucks et al. (1981). In contrast, the method we describe next produces globally optimal solutions, presuming that we approximate cost functions to the desired degree of precision. The methodology is the joint creation of myself and Clive Dym of Harvey Mudd College.

This second method, we indicated before, solves a slightly different problem than the first method. The first method assumed the reservoirs were in place and the capacities known. Also, the monthly contributions of individual reservoirs could change through the year. In contrast, this second method assumes capacities are not yet known but are to be determined so that the cost to deliver the firm system supply is minimum. In addition, the monthly contributions of individual reservoirs do not change through the year.

We indicated early in this chapter how the variables in Equation (3-8)

$$\alpha[h_o + (m/2)s_{t-1} + (m/2)s_t]x_t \geq d \tag{3-8}$$

could be separated by division by x_t. We also showed that this step could be followed by a piecewise linear approximation of the convex term d/x_t. The piecewise approximation is displayed in Equations (3-11), (3-12), and (3-13). These equations plus (3-5), (3-6), and (3-7) form a coherent constraint set over which some objective can be optimized. If capacity c in Equation (3-6) is treated as an unknown, then capacity can be minimized subject to these constraints.

As we pointed earlier, we can then iterate over a number of values of firm energy d, and a graph of capacity versus firm energy can be created. We called this a storage-energy curve. This curve is, however, a theoretical as opposed to practical trade-off between capacity and firm energy. It is a theoretical trade-off because the streamflow values that were utilized in the optimization are drawn from a hydrologic record that is past and will never repeat. Those flows constitute a representative history of flows only. To approach a practical curve of capacity versus firm energy we used the idea of deriving a rule curve for storage, a set of 12 end-of-month storage values that the reservoir operator would strive to achieve year after year. That is, 12 storage values, s_k, replace n storage values s_t.

Using only 12 variables for storage rather than n, as in Equations (3-16), (3-17), and (3-18), we can once again create a graph of capacity versus firm energy except that this graph represents a more practical level of energy output than the preceding graph. The graph is likely to have the characteristic shape of declining marginal increases in firm energy with each added unit of capacity; that is, the curve of firm energy versus capacity is likely to be concave. Of course, the curve of capacity versus firm energy is then convex, and this is the curve we will use to proceed. Such a curve of capacity versus firm energy is created for each of the multiple reservoirs in parallel.

For each reservoir site, the cost of building the reservoir to a particular capacity can be calculated. Hence, each of the capacity-versus-firm energy curves can be transformed into a cost-versus-firm energy curve.

To proceed, we need new notation. We let

$f_i(d_i)$ = cost to build reservoir i large enough to provide a firm supply of d_i

The shape of $f_i(d_i)$ will be site-specific and we cannot say in advance what its shape will be.

The problem we want to solve is one of determining the capacities of each of the three reservoirs that minimizes the system cost subject to the delivery of a firm system energy D, that is,

$$\text{Minimize} \qquad z = \sum_{i=1}^{3} f_i(d_i)$$

$$\text{s.t.} \qquad \sum_{i=1}^{3} d_i = D$$

We immediately reorganize this problem statement as amenable to solution by dynamic programming and hence, in the fashion of the mathematician, declare "there is a solution." The solution will be a firm energy and a cost and capacity for each of the three reservoirs, but for each reservoir, there will also be a storage-rule curve previously determined by solution of the problem of minimum capacity to deliver a firm energy.

The problem that we have structured, however, has a built-in assumption about the operation of each of the reservoirs, namely, that the reservoirs will not be operated in concert but independently of one another. Each reservoir i produces firm energy d_i each and every month of the year. As a consequence, a better solution might exist in which the reservoirs' contributions toward the firm system supply D varied with the month; in one month, a particular reservoir provides more than in another.

BIBLIOGRAPHY

Grygier, J. C., and J. R. Stedinger. 1985. "Algorithms for Optimizing Hydropower System Operation." *Water Resources Research,* Vol. 21, No. 1, pp. 1–11.

Loucks, D., J. Stedinger, and D. Haith. 1981. *Water Resource System Planning and and Analysis.* Prentice Hall, Englewood Cliffs, NJ.

Major, D., and R. Lenton. 1979. *Applied Water Resource System Planning.* Prentice Hall, Englewood Cliffs, NJ.

Pereira, M., and L. Pinto. 1985. "Stochastic Optimization of Multi-reservoir HydroElectric System." *Water Resources Research,* Vol. 21, No. 6, pp. 779–792.

Reznicek, K., and S. Siminovic. 1990. "An Improved Algorithm for Hydropower Optimization." *Water Resources Research,* Vol. 26, No.2, pp. 189–198.

Rosenthal, R. E. 1981. "A Nonlinear Network Flow Algorithm for Maximization of Benefits in a Hydroelectric Power System." *Operational Research,* Vol. 29, No. 4, pp. 763–786.

Trezos, T. and W. Yeh. 1987. "Use of Stochastic Dynamic Programming for Reservoir Management." *Water Resources Research,* Vol. 23, No. 6, pp. 983–996.

Turgeon, A. 1981. "Optimum Short-Term Hydro Scheduling from the Principle of Progressive Optimality." *Water Resources Research,* Vol. 17, No. 3, pp. 481–486.

Yeh, W., L. Becker, and W. Chu. 1979. "Real-Time Hourly Reservoir Operation." *Journal of Water Resources Planning and Management (ASCE),* Vol. 105 (WR2), pp. 187–203.

CHAPTER 4

COST ALLOCATION AND STAGING/SEQUENCING

COST ALLOCATION IN WATER SUPPLY

We begin with a problem in which a number of communities are together considering how to obtain stable water supplies. One can conceive of several situations relative to multiple communities seeking a stable supply of water. A number of reservoir sites may be available. A number of systems designs that include reservoir designs, pipeline networks, and treatment facilities may be possible. Of course, for any pair, triplet, quadruplet, and so on, of communities, only the best option need be considered. The communities may have the option of acting entirely alone, of joining in a single administrative unit (a water supply authority), or of forming subgroups or coalitions whose separate supplies individually provide the total water supply required by the multiple communities.

The choices the communities ought to make from an economic perspective are the choices that provide the needed quantities of water to all parties at the lowest total cost. Of course, communities may view such choices with other than economic lenses. It may often be true that a community's desire to remain separate and sovereign in as many matters as possible is so compelling that the community will choose an option that costs more, but ensures its independence to make its own decisions. Because communities exhibit this propensity to choose the independent option, researchers have explored options for formulas for allocating the costs to participants in regional efforts. The various formula options for cost allocation, including the popular method known as separable costs remaining benefits (SCRB), can all fail (Giglio and Wrightington, 1972). By "failure" is meant that costs are allocated in such a way that some grouping of participants can decide to "calve off" into a subcoalition (including a subcoalition of one) and do better in terms of the costs that will be incurred. That the formula allocations can fail does not meant that they will inevitably fail in each situation, but that the potential exists for failure.

One nonformulaic method is known that can be counted on to produce allocations that make all inferior subcoalitions nonattractive to the system participants, or at least can make participants indifferent to inferior subcoalitions. The method utilizes mathematical programming to determine a set of cost allocations such that any subgrouping of participants that is not part of the least cost arrangement of participants will not appear more attractive than the least cost arrangement (see Giglio and Wrightington, 1972; Heaney, 1997). The preceding sentences are very precise. They do not say that a subgrouping that is not part of the least cost arrangement will be unattractive. Instead, the language implies that cost can be allocated by this method in a way that will at least make participants indifferent between the best arrangement of participants and any other arrangement. To fully understand the language of indifference, we need to provide the actual formulation.

The mathematical programming formulation for cost allocation draws on the literature of n-person bargaining games. We illustrate the formulation with a three-community water supply example. Although we first illustrate cost allocation in the case of communities and their possible regional groupings, the methodology can be applied as well to the allocation of costs to various project purposes for a single reservoir providing multiple services; see Young et al. 1985. We will provide an example of this latter situation shortly.

Assume that three communities, A, B, and C, have the potential to develop (1) a common water supply (denoted ABC) that serves all three (the grand coalition), or (2) a supply that serves A and B with C furnishing its own separate supply, (3) a supply that serves B and C with A furnishing its own supply, (4) a supply that serves A and C with B furnishing its own supply, and (5) three separate supplies. These possibilities are illustrated in Figure 4-1, in which the circled letters represent the individual communities and the letters surrounded by a square represent the coalitions of these communities.

C_A = cost of A providing its own supply
C_B = cost of B providing its own supply
C_C = cost of C providing its own supply
C_{AB} = cost of A and B developing a joint supply
C_{BC} = cost of B and C developing a joint supply
C_{AC} = cost of A and C developing a joint supply
C_{ABC} = cost of A, B, and C developing a common supply
S_{AB} = $C_A + C_B - C_{AB}$ = savings available to A and B if they operate
     jointly rather than individually
S_{BC} = $C_B + C_C - C_{BC}$ = savings available to B and C if they operate
     jointly rather than individually

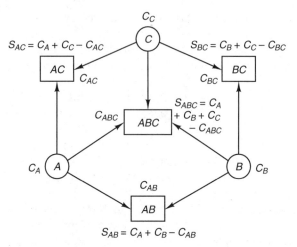

FIGURE 4-1 Alternate system configurations of communities A, B, and C with costs and savings.

$S_{AC} = C_A + C_C - C_{AC}$ = savings available to A and C if they operate jointly rather than individually

$S_{ABC} = C_A + C_B + C_C - C_{ABC}$ = savings available to A, B, and C if they choose to participate in a three-way effort

All these costs and savings are illustrated in Figure 4-1 next to the appropriate circle or rectangle. In the system shown in the figure, we assume that the cost of the grand coalition (A, B, and C) is less than that of any other configuration; that is,

$$C_{ABC} < C_A + C_B + C_C$$

$$C_{ABC} < C_{AB} + C_C$$

$$C_{ABC} < C_{BC} + C_A$$

$$C_{ABC} < C_{AC} + C_B$$

As a consequence of these cost relations, the constraints we will develop should make the participants feel that the grand coalition is at least as good as any other configuration [(AB and C), (BC and A), (AC and B)], and it is hoped better than some. To achieve this condition, cost shares need to be carefully allocated to each of the communities from go-it-alone costs of each of the participants. The actual cost share then is the go-it-alone cost less the allocated discount.

We define three unknowns:

X_A = dollar value of the discount offered to community A
X_B = dollar value of the discount offered to community B
X_C = dollar value of the discount offered to community C

The actual costs to the three communities then will be $C_A - X_A$, $C_B - X_B$, and $C_C - X_C$, and these costs will be calculated once the discounts have been determined.

What should the discounts be? Or, more precisely, what relations must the discounts satisfy?

First of all, the discounts allocated to any pair of participants must ensure that the savings they receive in aggregate by the allocation are at least as large as the savings they would receive if they separated from the grand coalition to form a coalition of two, with the third going it alone. For this three-participant problem, there are three pairs of two and hence three constraints:

$$X_A + X_B \quad\quad \geq S_{AB} \tag{4-1}$$

$$X_B + X_C \geq S_{BC} \tag{4-2}$$

$$X_A + \quad\quad X_C \geq S_{AC} \tag{4-3}$$

If the problem had four participants, and the grand coalition were still the optimal arrangement, there would be six such constraints for the six pairs of two, but there would be, in addition, four constraints for the four pairs of three.

Returning to the present situation, however, we see that the choice of values for X_A, X_B, and X_C that we need to make because of (4-1), (4-2), and (4-3) will cause each of the potential participants in each of the subcoalitions to be at least indifferent between the subcoalition and the grand coalition.

For example, A and B are allocated discounts (X_A and X_B to encourage them to participate in the grand coalition), and the discounts sum to a number at least as large as the savings that A and B would have obtained if they had chosen to form an AB partnership—rather than become participants in the grand coalition. Similarly, B and C are allocated discounts that sum to a number at least as large as the savings they would have acheived if they had chosen the BC partnership. And a similar situation obtains for A and C. Before we optimize some objectives, given these constraints, we explore more fully what these constraints may achieve.

Now the reader can see that it may be possible when the three constraints are met, to allocate discounts to A, B, and C in such a way that the communities are only made indifferent between the grand coalition and the more local partnerships, AB, BC, and AC. This would occur if all three constraints are met precisely as equalities. Even if only one of the constraints is met as an equality, say, for the AB partnership, A and B would be indifferent between the choosing the AB coalition and the grand coalition. Whatever objective is superimposed over these three constraints, the possibility of indifference is real. And if any pair of the participants are indifferent, then the program of discounts may fail to achieve the desired behavior. That is, the members of that participant pair may choose their local subcoalition rather than the grand coalition.

It follows that the three constraints are necessary to achieve indifference, but may not be sufficient to prevent subcoalitions from forming—especially if there are other perceived inequities. To correct this possible defect the right-hand side of each of the three constraints can be augmented by a small number, ε, perhaps 1/2 to 1% or more of the initial right-hand value. When these new constraints are satisfied, no comunity pair should be indifferent between their subcoalition and the grand coalition since they will be allocated savings that exceed the savings they could obtain in the more limited partnership.

Many solutions are typically possible for the set of constraints, so the issue is what criterion to apply to choose among the many possible discount allocation sets. Presumably a set of discount allocations does exist, which makes the grand coalition still feasible economically, given the other constraints. That is, the cost shares allocated (go-it-alone cost less discounts) are greater than or equal to the cost of the grand coalition:

$$C_A + C_B + C_C - (X_A + X_B + X_C) \geq C_{ABC} \qquad (4\text{-}4)$$

or

$$X_A + X_B + X_C \leq S_{ABC}$$

If the sum of discounts needed were greater than S_{ABC}, the ABC coalition would not appear to be optimal, that is, would not appear to be economically feasible, and should not be encouraged. Given this condition, it makes sense to ask what least sum of discounts might be given that still honors the three basic constraints relative to the subcoalitions. The objective and constraints then would be written

$$\text{Minimize} \quad Z \quad = X_A + X_B + X_C \qquad (4\text{-}5)$$

$$\text{s.t.} \quad X_A + X_B \quad \geq S_{AB} + \varepsilon \qquad (4\text{-}6)$$

$$X_B + X_C \geq S_{BC} + \varepsilon \qquad (4\text{-}7)$$

$$X_A \quad + X_C \geq S_{AC} + \varepsilon \qquad (4\text{-}8)$$

$$X_A, X_B, X_C \geq 0 \text{ and } \varepsilon = \text{a small number}$$

If the minimal value z of the allocated discounts sum to a number less than S_{ABC}, then the logical conclusion is that less discounts need to be given to make the participants see the grand coalition as attractive than the savings that the grand coalition provides. This may be an advantageous state of affairs; that is, one that can be exploited to the advantage of society (the group of participants). For instance, participants might be made to see the grand coalition as the preferred economic alternative even if they are given cost shares (go-it-alone minus allocated discount) that actually sum to a number greater than the cost of the grand coalition.

Several possibilities then occur. First, more savings could be allocated than the least sum of X_A, X_B, and X_C, of course, maintaining the constraints that are intended to "block" inferior subcoalitions. A logical question we shall explore shortly is under what criterion should additional discounts beyond the minimum sum be allocated?

Second, the discounts corresponding to the least sum of X_A, X_B, and X_C could actually be collected and the excess $[S_{ABC} - (X_A + X_B + X_C)]$ could be used by the authority to cover other costs, such as administrative costs, or to improve the system in other ways. For future reference, we call these the "excess savings." The money could even be banked and set aside for future expansion of the system to accommodate anticipated growth. Another alternative "conceptually" is to collect all of the allocated costs and return the excess to the participants via some formula. Such a step would need to be taken cautiously so that the basic constraints that make participants indifferent to inferior subcoalitions remain in force. In fact, returning the excess in an intelligent way may provide an opportunity to further solidify the "attractiveness" of the superior alternative. We next examine how this could be the case.

"Sovereignty" is an issue against which advocates of regionalization consistently struggle. Sovereignty may be defined for our purposes as the urge for communities to remain separate in all possible ways—in education, transportation, water supply, and so on. Communities have been known to operate/behave in ways that appear to be detrimental to their economic interests in order to preserve their sovereignty or because they distrust other communities. The behavior could well be manifest in not joining a regional effort such as the formation of a joint water supply or a solid waste district.

Such behavior could take place even when costs are allocated in a way that makes all inferior subcoalitions actually unattractive. Your immediate reaction could well be "That's irrational," and you would be right. Although the assumption of rational economic behavior is at the root of the formulation we have provided, not all communities always behave rationally. The desire for the freedom to operate independently may override pocketbook issues. Thus, we have to admit that the formulation, as it appears so far, makes the assumption of rational economic behavior. Is there hope to make the formulation work in *all* cases? Probably not. Irrationality is just what it says. No amount of rational information can "cure" irrationality in every case. Nonetheless, it may be possible to so arrange the balance sheet that "homo economicus" prevails more frequently in the contest.

This is where the situation in which the minimum total discounts less than the savings provided by the grand coalition can be used to effect. Since all of the grand coalition savings can be distributed, more can be given back than the minimum without "breaking the budget." And the manner in which the additional discounts are allocated, as long as the group rationality and individual rationality constraints are not violated, can alleviate some concerns on equity.

Equity can have many dimensions. We will focus on just one for this discussion—the evenness of the fractional discounts from the go-it-alone costs. (See Ratick, Cohon, and ReVelle, 1981.) For three participants, the three fractional discounts are X_A/C_A, X_B/C_B, and X_C/C_C. If the minimum discount subject to rationality constraints yields an allocation in which one participant has a 15% discount and two have a 10% discount, then, despite the appearance of a rational allocation, the two participants with the 10% discounts may be tempted to withdraw from cooperation because of an "unfair" allocation. This is the case in which the distribution of the excess savings can be utilized to good effect.

For a given level of total discounts given (up to S_{ABC}), we can minimize the maximum difference between fractional discounts, subect to the rationality constraints. In nonstandard form, this is

$$\text{Minimize} \quad z \quad = U - L \tag{4-9}$$

$$\text{s.t.} \quad X_A/C_A \leq U \tag{4-10}$$

$$X_B/C_B \leq U \tag{4-11}$$

$$X_C/C_C \leq U \tag{4-12}$$

$$X_A/C_A \geq L \tag{4-13}$$

$$X_B/C_B \geq L \tag{4-14}$$

$$X_C/C_C \geq L \tag{4-15}$$

$$X_A + X_B \geq S_{AB} + \varepsilon \tag{4-16}$$

$$X_B + X_C \geq S_{BC} + \varepsilon \tag{4-17}$$

$$X_A + X_C \geq S_{AC} + \varepsilon \tag{4-18}$$

$$X_A + X_B + X_C \leq M \tag{4-19}$$

$$X_A, X_B, X_C, U_L \geq O$$

Where the new notation is

U = largest fractional discount given
L = smallest fractional discount given
M = limit on total discounts given ($M < S_{ABC}$)

The larger M is allowed to be, the smaller can be the maximum difference in fractional discounts. If equality can be achieved in fractional discounts without giving back more than S_{ABC}, then equity in at least one dimension has been achieved.

If the least sum of dollar discounts is less than S_{ABC}, then we can conceive of still another way to "fight" indifference. The ε that was added to the right-hand side of (4-6), (4-7), and (4-8) was a small number. If it could be increased from ε to E, a more substantial number, then the indifference of simply matching savings might be overcome.

The mathematical program would be written (in nonstandard form)

$$\text{Maximize} \quad z = E$$

$$\text{s.t.} \quad X_A + X_B \qquad \geq S_{AB} + E$$
$$X_B + X_C \geq S_{BC} + E$$
$$X_A + \qquad X_C \geq S_{AC} + E$$
$$X_A + X_B + X_C \leq M$$

where M is an upper limit on the discounts that can be given back and is less than S_{ABC}. The larger the value of M, the larger E can be made and the more likely that indifference will not occur.

The model could also utilize a single factor F that multiplies each of the savings, maximizing that factor as a means of inducing a strong blocking effect.

$$\text{Maximize} \quad z = F$$
$$\text{s.t.} \quad X_A + X_B \qquad \geq F * S_{AB}$$
$$X_B + X_C \geq F * S_{BC}$$
$$X_A + \qquad X_C \geq F * S_{AC}$$
$$X_A + X_B + X_C \leq M$$

Other models are possible as well.

We will leave the subject of cost allocation between communities for the provision of water supplies to proceed to a discussion of another sort of cost allocation in water resources. Nonetheless, we note that the subject is still capable of extension, that other measurers of equity are all possible—and optimizable and that other objectives can increase the attractiveness of the grand coalition. We note, too, that regional solutions for water supply may be better than local provision in another way than economically. Regional (larger) water systems may be better maintained and produce water of better and more consistent quality than highly localized systems. Furthermore, system failures occur less frequently. So the regional solution, even at the expense of sovereignty, has other benefits than economic.

COST ALLOCATION IN MULTIPLE-PURPOSE RESERVOIRS

The idea of allocating costs of joint projects extends as well to single projects with multiple purposes. In this case, it is the purpose that is allocated a cost rather than the participating community. The example is a reservoir that is used for four purposes (see "Cost Allocation: Methods, Principles, Applications"). The purposes may be water supply (A), flood control (B), navigation (C), and hydropower (D). Each purpose could be served by a reservoir of a particular size or cost. A reservoir serving the water supply function only costs C_A, the reservoir serving only flood control, C_B, the reservoir serving only navigation, C_C, and the reservoir serving only the hydropower purpose, C_D. Different combinations of purposes (see Table 4-1) require reservoir capacities that have known costs. It is known in advance that the reservoir will be built to supply four functions.

In this situation, it seems easier to think in costs allocated rather than in discounts allocated. We let Y_A, Y_B, Y_C, and Y_D be the costs allocated to each of the four purposes. By using costs rather than discounts, the constraints (Table 4-2) are of the less than or equal to form, with the exception of the last constraint. The last constraint requires that the reservoir with all of its four functions be fully paid for. It is likely that there are many solutions to these constraints. We might choose that set of allocations that causes each purpose to pay, as nearly as possible, the same fraction of the cost of providing that purpose alone. To operationalize this idea, we would minimize the maximum difference between the largest and the smallest fractional payment. That

TABLE 4-1   Cost of Achieving Varying Combinations of Purposes

Purpose	Water Supply	Flood Control	Navigation	Hydropower	Cost
	√				C_A
		√			C_B
			√		C_C
				√	C_D
	√	√			C_{AB}
	√		√		C_{AC}
	√			√	C_{AD}
		√	√		C_{BC}
		√		√	C_{BD}
			√	√	C_{CD}
	√	√	√		C_{ABC}
	√	√		√	C_{ABD}
		√	√	√	C_{BCD}
	√		√	√	C_{ACD}
	√	√	√	√	C_{ABCD}

TABLE 4-2 Cost Allocation Constraints

Y_A				$\leq C_A$	(4-20)
	Y_B			$\leq C_B$	(4-21)
		Y_C		$\leq C_C$	(4-22)
			Y_D	$\leq C_D$	(4-23)
Y_A	$+ Y_B$			$\leq C_{AB}$	(4-24)
Y_A	$+$	Y_C		$\leq C_{AC}$	(4-25)
Y_A	$+$		Y_D	$\leq C_{AD}$	(4-26)
	Y_B	$+ Y_C$		$\leq C_{BC}$	(4-27)
	Y_B	$+$	Y_D	$\leq C_{BD}$	(4-28)
		Y_C	$+ Y_D$	$\leq C_{CD}$	(4-29)
Y_A	$+ Y_B$	$+ Y_C$		$\leq C_{ABC}$	(4-30)
Y_A	$+ Y_B$	$+$	Y_D	$\leq C_{ABD}$	(4-31)
	Y_B	$+ Y_C$	$+ Y_D$	$\leq C_{BCD}$	(4-32)
Y_A	$+$	Y_C	$+ Y_D$	$\leq C_{ACD}$	(4-33)
Y_A	$+ Y_B$	$+ Y_C$	$+ Y_D$	$= C_{ABCD}$	(4-34)

is, we would add the following objective and constraints to the constraints in Table 4-2:

$$\text{Minimize} \quad Z = U - L \tag{4-35}$$

$$\text{s.t.} \quad Y_A/C_A \leq U \tag{4-36}$$

$$Y_B/C_B \leq U \tag{4-37}$$

$$Y_C/C_C \leq U \tag{4-38}$$

$$Y_D/C_D \leq U \tag{4-39}$$

$$Y_A/C_A \geq L \tag{4-40}$$

$$Y_B/C_B \geq L \tag{4-41}$$

$$Y_C/C_C \geq L \tag{4-42}$$

$$Y_D/C_D \geq L \tag{4-43}$$

Constraints (4-36) to (4-43) plus (4-20) to (4-34) and the objective (4-35) constitute one of several possible programs to allocate costs across the purposes of the reservoir.

TIME STAGING OF WATER SUPPLY ALTERNATIVES

The practice of planning for the water supply of a community takes account of projections of growth and demand as well as the costs of alternatives. In a world of perfect information, planners would also have information on the response of demand to price and would take account of the "demand curve" in their planning of supplies. Unfortunately, most planning agencies do not possess information on demand curves even for short excursions into the future. Thus, planning for physical water supply alternatives will rarely, if ever, take account of demand curves.

Instead, growth in population equivalents may be projected and the demand corresponding to that growth calculated at either a specific point in time or at a number of points in time. Planning is then (A) focused on that specific future date or (B) considers the entire time line of demand growth. The first situation, focus on a specific date and demand, is more conventionally utilized. It is the situation we have described up till now, in which water supply planners ask the question: "What is the smallest reservoir needed to supply a demand of q units of water per day?" The demand q may be the projected usage 10 or 20 years hence.

The second situation, however, is the one that commands our attention in this chapter. The question here is: "Which alternatives should be built and when should they be built so that demand throughout the horizon of concern is met at the least total cost?" Our assumption will be that a curve of demand versus time can be supplied for the planning process. The approach here will be first to formulate the problem as a zero-one programming problem and describe the solution properties of the formulation. Then, we will develop an exact and immediate procedure for the problem when demand grows linearly in time and suggest the procedure as a heuristic when demand is not growing linearly. We assume an inflation-free enviornment.

The problem is one of determining the date at which each project should be available to come into service. If a project needs to be complete by the beginning of year 10 and takes three years to complete, it must be begun by the start of year 7. At the time that the project is about to go into service, demand will be just reaching previously available capacity. As soon as the project is in service, an immediate overcapacity will have been created, with overcapacity momentarily equal to the yield of the project coming on line, but declining through time as demand grows. At the moment this new overcapacity has declined to zero, the next project must be available to come on line. Figure 4-2 illustrates the concept.

A Zero-One Programming Formulation

To formulate the problem, we introduce the following notation for parameters and indices:

m, i = number and index of projects, respectively
n = number of instances (years) at which demand is to be met through time
r = interest rate
e_i = yield of project i

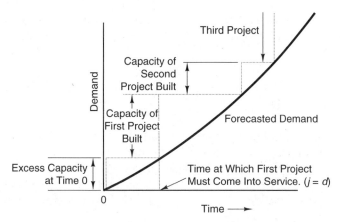

FIGURE 4-2   Demand and capacity time lines.

c_i = today's cost of project i
c_{ij} = cost of project i built at time j in the future:[1] $c_i/(1 + r)^j$
d_j = demand at time j

The general decision variable is

x_{ij} = 1, 0; it is 1 if project i is built at time j, and 0 otherwise
y_i = 1, 0; it is 1 if project i is *not* built during the horizon, and 0 otherwise

The problem is to schedule these projects so that demand is met at each and every point in time and the cost is least. Two types of constraints are needed. The first says that a project can be built at most once over the horizon. If it is not built, variable y_i must be one.

$$\sum_{j=1}^{n} x_{ij} \leq 1 \qquad i = 1, 2, \ldots, m \tag{4-44}$$

This constraint can be converted to an equality by the addition of the slack variable y_i to the left-hand side:

$$\sum_{j=1}^{n} x_{ij} + y_i = 1 \qquad i = 1, 2, \ldots, m$$

Although it is not strictly necessary to define and add the slack variable y_i, we will see later that the variable provides us useful versatility, allowing easy modification of the problem when projects are of a divisible nature.

---

[1] The money needed to be set aside today to have c_i, at time j.

The second constraint says that at each reference point in the future, the sum of yields of projects built must equal or exceed the demand at the reference time.

$$\sum_{i=1}^{m} e_i \sum_{t=1}^{j} x_{it} \geq d_j \qquad j = 1, 2, \ldots, n \tag{4-45}$$

If project i has been built by time j (including at time j), the interior sum is 1 and the yield of project i can be counted toward meeting the demand at j. Presumably, the first project to be built can be completed in the interval between the present moment and the time at which current capacity needs to be augmented by additional capacity (time $j = 1$). The objective to be optimized is

$$\text{Minimize } z = \sum_{i=1}^{m} \sum_{j=1}^{n} c_{ij} x_{ij} \tag{4-46}$$

In total, the problem has $mn + m$ variables and $m + n$ constraints.

This formulation has appeared in Major and Lenton (1979) and Knudsen and Rosbjerg (1977), although the latter recommended a dynamic programming algorithm along the lines of Morin and Esogbue (1971) or of Butcher et al. (1969). A specially coded algorithm seems unnecessary, however, in the present era, given the solution properties of the linear programming problem in which the zero-one requirements are relaxed to simple nonnegativity.

The problem described by objective (4-46), and constraints (4-45) and (4-44), can be seen to be of a special form in which the number of fractional variables in a linear programming solution (read "extreme point") is strictly limited. We first note that, for any linear programming problem, the number of positive variables at an extreme point solution is no more than the number of structural constraints. This problem has $m + n$ constraints, m defining when or if each project is built and n requiring that demand be satisfied. Thus, we know first that no more than $m + n$ variables will be positive—despite the fact that the problem has $mn + m$ variables. The constraints

$$\sum_{j=1}^{n} x_{ij} + y_i = 1 \qquad i = 1, 2, \ldots, n \tag{4-47}$$

are written one for each project. It can be seen that, for each such constraint, at least one variable from the constraint, and possibly more, must be positive to satisfy the relation. Given that there are m projects and m such constraints, we have already used up m of the $m + n$ positive variables that characterize an extreme point. That leaves us n variables (the number of reference times) to distribute amount m project constraints. If there are 100 possible project alternatives (m) and 25 years in which demand is to be met (n), then no more than 25 projects could be "split" in any fractional solution.

Solution of the integer program will likely require branch and bound applied repeatedly to the linear programming solution. Branch and bound may be guided in a number of ways, including the specification of the variables on which to branch and the order of branching on these variables. Probably the variables on which to branch first would be the y_i and the order of branching would be based on the ratio of e_i/c_i. Since y_i equal to one means project i is never built, it would seem efficient to branch first on that i with the least yield-to-cost ratio.

The problem described by objective (4-46) and constraints (4-45) and (4-47) can be modified to handle the situation in which one or more of the projects can potentially be subdivided into distinct phases or stages. Perhaps a dam project on a particular river might be built in one or two steps. It could be built first to one height and later extended to a second height, if it were economicaly attractive to do so. Alternatively, it could be built in a single step to the larger size at a cost lower than the cost of the two steps taken separately. The reason that building in two steps could be attractive is that postponing costs until they are needed provides some savings.

Suppose one of the projects in which the dam may be built to the larger size in a single step is labeled as A. For this project, the two sequential steps that ultimately produce a dam of the larger size are called B and C, where project B is built first and C is built second. Needed parameters are the yields e_A, e_B, and e_C (where $e_A = e_B + e_C$) and the costs C_A, C_B, and C_C (where $C_A < C_B + C_C$). The alternatives A, B, and C are among the m alternatives under consideration, but must be treated in a special fashion for two reasons. First, project C cannot be built until project B is already completed. Second, if B is undertaken, then project A cannot be undertaken. (It follows, of course, that since C cannot be built unless B is, that enforcement of the second condition will ensure A and C will not both be undertaken.) These two requirements can be enforced with two sets of constraints that use the the slack variables y_A, y_B, and y_C. Recall that any y_j equaling one means that the particular project is not undertaken. The negation of some y_j, that is, $(1 - y_j)$ equal to one means that the project is undertaken.

The first condition, in which C cannot be begun unless B is completed, can be enforced by the constraint

$$1 - y_C \leq 1 - y_B \qquad (4\text{-}48)$$

which is simplified to

$$y_B \leq y_C \qquad (4\text{-}49)$$

The second condition, in which project B and project A cannot both be undertaken, can be modeled by the constraints

$$1 - y_B + 1 - y_A \leq 1$$

or

$$y_A + y_B \geq 1 \tag{4-50}$$

This second version of the constraint can be read as saying that at least one of the two projects, A and B, *cannot* be built—which is, of course, exactly what is desired.

These projects, A, B, and C, would be in the list of possible projects with their costs in the objective and yields in the constraints that require sufficient water at the appropriate temporal reference points. They would, however, be treated with these special constraints described before in addition to the other constraints.

The reader may be wondering if the staging or ordering rules might be violated since we have not utilized time in the divisible project constraints. The requirement that at least one of the projects A and B cannot be built has no temporal character whatsoever. So no order of staging problem could be created here. However, the requirement that C cannot be undertaken unless B is built does imply that B will be built first. Thus, it would appear that the constraint $y_B \leq y_C$ *could be* insufficient to enforce the temporal requirement that B is built first in time. Although this would be an unlikely occurrence because the ratio of cost to yield is likely to be higher for project C, it could occur. To prevent such an occurrence, we could utilize a more completely specified set of constraints of the form

$$x_{Cj} \leq \sum_{t=1}^{j-1} x_{Bt} \qquad j = 2, 3, \ldots, n \tag{4-51}$$

which says that project C cannot come into service in any period j unless project B has been built in some prior period.

An Exact Algorithm for Staging Under Linear Growth

This unpublished algorithm due to Sheer (1974) provides the order of the projects to build and, with a negligible amount of additional calculation, their times of coming on line. We use the notation established earlier with the addition of parameter g. Coefficient g is the demand growth per year, for instance, millions of gallons/day increase per year. This number does not change through the planning horizon.

A particular project supplies a yield e. Given that demand growth each year is g, that project will, once put into service, provide sufficient water for e/g years, after which a new project must be put on line.

We begin by comparing two projects, A and B, and the two possible orders of their establishment: A then B and B then A. Suppose the projects are built in the order A then B. The cost is

$$C_A + \frac{C_B}{(1 + r)^{e_A/g}}$$

where the first term of the sum is the cost of building project A today and the second term of this sum is the cost of building B at e_A/g years after project A comes into service, that is, when project A has become insufficient to meet further growth in demand. The second term is the present value of building project B at e_A/g years in the future.

If the order of building is B then A, the cost is

$$C_B + \frac{C_A}{(1 + r)^{e_B/g}}$$

where the first term of the sum is the cost of building project B today and the second term of this sum is the cost of building A at e_B/g years after project B comes into service, that is, when project B has become insufficient to meet further growth in demand.

We let

$$t_A = e_A/g \qquad \text{and} \qquad t_B = e_B/g$$

These are the times that each project lasts before the next project must be available to meet the linearly growing demand.

We will assume that the sequence A then B costs less than the sequence B then A. That is,

$$C_A + \frac{C_B}{(1 + r)^{t_A}} < C_B + \frac{C_A}{(1 + r)^{t_B}} \qquad (4\text{-}52)$$

Manipulation of this inequality yields

$$\frac{C_A}{1 - (1 + r)^{-t_A}} < \frac{C_B}{1 - (1 + r)^{-t_B}} \qquad (4\text{-}53)$$

where the term on the left-hand side involves only the properties of project A, namely, its cost and yield, and the term on the right-hand side involves only the properties of project B. We can define the term on the left-hand side as R_A (rating of A) and the term on the right as R_B (rating of B). The two terms characterize the two projects in terms of the relative cost of their utilization in the sequence. Since the assumption that project order A then

B was least costly resulted in $R_A < R_B$, we can reason in reverse that, if R_A is less than R_B, then building project A then B is less costly then the reverse.

If three projects A, B, and C are least costly to build in the order A then B then C, we will find that $R_A < R_B < R_C$. And if $R_A < R_B < R_C$, the least cost project order is A, B, and C. The argument may be extended to any number of projects; that is, calculating R_j ($j = 1, 2, \ldots, m$) and ranking the projects from least R_j to the most costly R_j yields the optimal order of staging. Ties in R_j indicate that the tied projects can occupy any position between the smallest and next largest R_j—without changing the value of the objective function. This remarkably simple procedure provides an optimal solution to the sequencing of water supply alternatives under linear demand growth. No mathematical program, integer/combinatorial or dynamic, is needed. The procedure appears to have value, however, even beyond the confines of the growth scenario for which it was developed. Although the linear growth scenario is the only one for which the procedure finds the exact optimal solution, the procedure provides an effective heuristic for other demand growth functions.

Suppose growth proceeds according to some function other than linear. At the moment in time when some project must be in place to meet growing demand, we conceptually place in turn each project $j = 1, 2, \ldots, m$ into service. We then calculate or observe graphically the time that the project j lasts before its yield is just insufficient to meet the continually growing demand. This is the time that the *next* project must come into service. For project j, this time is t_j. Then t_j values are determined for all m projects. This t_j replaces its conceptual equivalent in linear demand growth, namely, e_j/g. For each project, a value can be calculated for R_j, namely,

$$R_j = \frac{C_j}{1 - (1 + r)^{-t_j}} \qquad (4\text{-}54)$$

The minimum R_j value is found from the m values, and the project with the minimum value is placed into service as the first project to be built in the sequence. This project now remains in service until its yield is just insufficient to meet growing demand at which time one of the remaining $(m - 1)$ projects must be available for service.

To decide which of these projects is selected next, each in turn is conceptually placed into service at the time the first project "runs out." Demand growth rates during the interval of the second project are, of course, different than during the first interval since this is not linear demand growth. For each of the $(m - 1)$ projects, a value of t_j is determined by calculation or observation; this is the time that each project will last before a third project must be placed in service. For each of the $(m - 1)$ projects, a value of R_j is now calculated. The project with the smallest R_j is chosen as the second project to be built in sequence. It lasts for a period that reflects the current rate of growth of demand and the project's capacity. At the moment that the second project becomes insufficient to meet growing demand, a third project from among the $(m - 2)$ remaining must be available. Hence, the process repeats once more with selection of the third project as the one with the least currently recalculated R_j from among the $(m - 2)$.

Sheer (1974) indicates that this near-term selection procedure is quite effective in finding the optimal solution for staging facilities. Indeed, the process of actual decision making—as opposed to the process of recommending an order for staging the facilities—ought to proceed incrementally. It ought to proceed with near-term decisions and with long-term plans because conditions change. New ideas may provide opportunities to utilize new alternatives. Development and competition from other communities may foreclose previously available options. Inflation may affect some projects differentially, and real interest rates may change. All these factors suggest strongly that decisions should be made in the near term, but that plans should be just plans—although purchases of land to lock in options are still appropriate.

Thus, only the near-term decisions that come from exact or heuristic methods would be expected to be implemented; there should be no risk of running out of water, but the remainder of a recommended plan should be viewed as flexible. And this suggests that, as long as the first choice in staging is very good, that recalculation of the second choice—as the time to implement a second choice approaches—is appropriate and desirable. Thus, a "greedy" procedure for choosing each new alternative matches the way decisions are actually made.

TIME STAGING AND COST ALLOCATION

This concluding section is necessary because leaving these two topics without placing them alongside one another in a more integrated form gives a false impression that these topics are placed in the same chapter for convenience only. That does not mean we can offer genuine progress in integrating the two topics here, but we can, at the least, begin to lay out the connections between them and, in the process perhaps, illuminate the modeler's art.

The successful modeler is one who slices the pie "just right," large enough to be able to attack an important and realistic problem, but small enough that the problem can be solved successfully. But there is room, too, to envision the problems in need of solution even if they are beyond the reach of today's tools and our imagination for structuring. Cost allocation and time staging are individually two edible slices of pie. Together, they cannot, at least at present, be swallowed and/or digested. Nonetheless, the reader should know that the two problems are really tightly connected. We can offer only a verbal description of the problem and leave the mathematical formulation to others with the talent and tenacity to structure it.

Envision a cluster of communities, each community with a growing demand. There are a number of physical alternatives for supplying the needs of these communities through time. The individual communities each have access to some subset of the alternatives. As a single group, the communities have access to all of the alternatives. We will assume that through time, the grand coalition, staged in time, remains the best strategy for all of the communities. The question is how the communities should share the costs of the meeting their needs and which alternatives should be built and when.

BIBLIOGRAPHY

Cost Allocation

Giglio, R., and R. Wrightington. 1972. "Methods for Apportioning Costs Among Participants in Regional Systems." *Water Resources Research,* Vol. 8, pp 1133–1144.

Heaney, J. 1997. "Cost Allocation in Water Resources." In *Design and Operation of Civil and Environmental Engineering Systems,* C. ReVelle and A. McGarity, Eds. John Wiley, New York.

Ratick, S., J. Cohon, and C. Revelle. 1981. "Multi-objective Programming Solutions to N-person Bargaining Games." In *Organizations: Multiple Agents with Multiple Criteria,* J. Morse, Ed. Number 190 in Lecture Notes in Economics and Mathematical Systems. Springer-Verlag.

Young, P., Ed. 1985. *Cost Allocation: Methods, Principles, Applications.* North Holland.

Staging/Sequencing

Butcher, W., Y. Haimes, and W. Hall. 1969. "Dynamic Programming for the Optimal Sequencing of Water Supply Projects." *Water Resources Research,* Vol 5, pp 1196–1204.

Knudsen, J., and D. Rosbjerg. 1977. "Optimal Scheduling of Water Supply Projects." *Nordic Hydrology,* Vol 8, pp 177–192.

Major, D., and R. Lenton. 1979. *Applied Water Resource Systems Planning.* Prentice Hall, Englewood Cliffs, N.J.

Morin, T., and E. Esogbue. 1971. "Some Efficient Dynamic Programming Algorithms for Optimal Sequencing and Scheduling of Water Supply Projects." *Water Resources Research,* Vol. 7, pp 479–484.

Sheer, D. 1974. "Economic Sequencing of Public Works Facilities." Ph.D. diss., Johns Hopkins University, Baltimore.

CHAPTER 5

INTEGRATING RESERVOIR SERVICES

To this point, we have shown how to model and optimize a number of basic reservoir functions. First, we examined the provision of a water supply by a single reservoir. We next showed how to incorporate flood control into water supply reservoir models. We did this by modifying the capacity limits of the water supply reservoir by month by subtracting the free storage volume that was needed for capturing floods in that month. We then investigated how the releases from multiple reservoirs on parallel streams could be merged into a common water supply. We next modeled hydropower production, structuring nonlinear constraints that required a specified level of firm power. We were able, by piecewise approximation, to convert the nonlinear constraints to linear constraints. Finally, we examined optimal time staging of reservoirs to meet a growing demand and we investigated how costs could be allocated among communities that entered into a joint venture to provide their water supply requirements. We also showed how cost allocation ideas could be applied as well to the division of the reservoir's cost among the functions of water supply, flood control, hydropower, and navigation.

The first function, water supply, is probably the most basic service the reservoir provides. Flood control is probably the second most basic service. Hydropower is both very important and the most difficult of the services to model, not only because of its inherent nonlinearity, but because of the various economic environments in which it is produced.

The principal reservoir services remaining are generally easier to model. One such remaining service is flow maintenance (1) for navigation or (2) for dilution of pollution or (3) for environmental purposes such as enhancement of conditions for fish spawning. Another service is recreation. Whereas flow maintenance implies sustaining the flow in the stream, possibly even through the dry season, recreation requires a relatively steady, low fluctuation storage volume in the reservoir. Neither of these services is particularly hard to model.

The last reservoir service we need to discuss is irrigation. Although superficially irrigation might seem to be very similar to water supply in its exercise, in fact, irrigation needs are often more flexible. Water supply implies use by cities for people and industry. Irrigation means application of water in agriculture. Water supplies are more firm, more precise, and failures to meet water supply needs are costly politically. Agricultural needs have targets with some "give" to them. In addition, a shortfall of irrigation water in one early season may decrease needs later because there are fewer crops to water.

Thus, to conclude our tour of reservoir modeling, we assign all the functions discussed earlier to a single reservoir. We merge into a single model: water supply, flood control, hydropower production, flow maintenance, recreation, and irrigation. And we show how to model these services simultaneously.

MODELING THE MULTIPURPOSE RESERVOIR

We begin by assuming we have a reservoir of capacity c and a long flow record to accompany it. From this long flow record, the worst drought is abstracted for our analysis. That is, we will look at how the reservoir is operated for these multiple functions in the worst water availability environment that been experienced.

We will initially need to add one more variable and a number of parameters to our lengthy set of notation. As we proceed, a few more variables will be needed. The new variables and parameters we need now are

z_k = unknown amount of water to be delivered for irrigation in month k ($k = 1, 2, \ldots, 12$)

t_k = given target or ideal irrigation delivery in month k ($k = 1, 2, \ldots, 12$)

q_k = known water supply need that must be provided in month k ($k = 1, 2, \ldots, 12$)

d_k = stated energy requirement for month k ($k = 1, 2, \ldots, 12$)

u_T = known maximum monthly flow through the turbines

f_k = specified streamflow maintenance need, month k ($k = 1, 2, \ldots, 12$)

The value of the irrigation target, t_k, may be larger during this period of low streamflows than during a period of normal flows. This condition may obtain because the time of low streamflow may reflect rainfall drought in the irrigation region as well as in the watershed, resulting in greater irrigation needs by crops. The reservoir is assumed in placed with a fixed capacity c.

The model needs some overarching purposes for us to provide an appropriate framework for modeling. Thus, we assume that the water supply amounts, q_k, must be provided. Additionally, the hydropower needs in month

k must be met, and finally the flood storage volumes, the v_k introduced earlier, are essential to maintain.

Our first modeling step is to restructure all the basic reservoir equations:

$$s_t = s_{t-1} + i_t - w_t - q_k - x_t - z_k \qquad t = 1, 2, \ldots, n;$$
$$k = t - 12 \left[(t - 1)/12 \right] \qquad (5\text{-}1)$$

$$s_t \leq c - v_k \qquad t = 1, 2, \ldots, n; \; k = t - 12[(t - 1)/12] \qquad (5\text{-}2)$$

$$s_n \geq s_o \qquad (5\text{-}3)$$

$$x_t \leq u_T \qquad t = 1, 2, \ldots, n \qquad (5\text{-}4)$$

$$d_k/x_t - (\alpha m/2)s_{t-1} - (\alpha m/2)s_t \leq \alpha h_o \qquad t = 1, 2, \ldots, n;$$
$$k = t - 12 \left[(t - 1)/12 \right] \qquad (5\text{-}5)$$

$$x_t + w_t \geq f_k \qquad t = 1, 2, \ldots, n; \; k = t - 12[(t - 1)/12] \qquad (5\text{-}6)$$

The mass balance equation (5-1) now contains all of its prior terms plus more removal terms. As before streamflow, i_t, enters during month t. As it was earlier, spill, w_t, is simply the result of an inability to store water because of capacity limitations or, in this case, an inability to use excess water by passing it through the turbines—because of a flow limitation through the turbines. The water supply requirement, q_k, is drawn off in each of the 12 months of the year; this quantity cannot be directed through the turbines. Release through the turbines to the stream, x_t, is used to generate power. Lastly, a quantity of water, z_k, is drawn off in each of the 12 months for irrigation. This is a busy reservoir.

In addition, constraint (5-2) limits the volume of water stored to the portion of the reservoir that can be devoted to water storage in month k, $c - v_k$. As we noted in Chapter 1, s_t, the end-of-period storage, should really be replaced by the average amount of water stored in the reservoir during month t. Constraint (5-3) prevents water from being borrowed. Constraints (5-4) limit flow through the turbines to u_T. Constraints (5-5) require that a power level d_k be provided in month k. The constraints are nonlinear, but each can be linearized by piecewise approximation, as described in Chapter 3. This step will be assumed. Constraints (5-6) are the flow maintenance requirements. These constraints may force spill to be positive in some months even when storage does not reach the storage limit $c - v_k$.

At this point, four services have been registered in the operational equations (5-1) to (5-6). These are water supply, flood control, hydropower production, and flow maintenance. The irrigation function is partially accounted by the z_k term in (5-1), but the recreation function remains completely unaccounted. It is not registered anywhere so far, but will be soon. The four services that have been fully modeled may be thought of as services with requirements that must be met.

Irrigation and recreation may be thought of as services that are needed but for which strict requirements are not applicable. Meeting these additional needs as well as possible, given that the other needs are met precisely, may be sufficient. The meeting needs "as well as possible" suggests that the measures of how well these needs are met can be placed in the objective function or can be traded off against one another, or, at least, are not "hard constraints."

We will treat the objective functions for irrigation and recreation services as ones that can be traded off against one another. For each of these services, several objectives are possible and plausible.

THE IRRIGATION OBJECTIVE

For irrigation, our first objective will be to minimize the maximum fractional deviation from the target for irrigation flow over all months. We will assume more water than the target will not be provided since water in excess of the target would not be regarded as beneficial and can be released to the stream—unless there are upper limits on stream flow. That is,

$$z_k \leq t_k \qquad k = 1, 2, \ldots, 12 \qquad (5\text{-}7)$$

We choose to minimize the maximum as the first objective because irrigation losses are probably convex with increasing shortfalls from the target. The fractional deviation from the irrigation target in month k is

$$\frac{t_k - z_k}{t_k}; \qquad k = 1, 2, \ldots, 12$$

It is this term that will minimize the maximum. Thus, we write

$$\frac{t_k - z_k}{t_k} \leq U \qquad k = 1, 2, \ldots, 12 \qquad (5\text{-}8)$$

where U is the unknown upper bound of fractional deviations (shortfalls) to be minimized. In standard form, this constraint becomes

$$t_k U + z_k \geq t_k \qquad k = 1, 2, \ldots, 12 \qquad (5\text{-}9)$$

and the objective would be to minimize U. We could also minimize the maximum shortfalls as opposed to the maximum fractional shortfall. In this case, the constraint would be, in nonstandard form,

$$t_k - z_k \leq U$$

Again, we would minimize U.

An interesting alternative would be to minimize the sum of monthly losses—if a loss function is available. Such a function would be expected to be convex and would achieve its minimum when the release to irrigation equals the target. That is part of the justification for the minimization of the maximum shortfall or minimization of the maximum fractional shortfall from the target.

Loss functions are problematic to determine. They depend on the irrigation infrastructure in place and the crops chosen, both of which are liable to change through time. Furthermore, the loss function (Northern Hemisphere) may be different in July from its anticipated form if, in June, a large shortfall from the target occurred. That is, a crop that has suffered from a shortage in June will not respond fully to irrigation releases in July. Thus, loss functions determined a priori are unlikely to provide the desired economic precision. Minimization of the sum of deviations or the squares of deviations from targets or minimization of the sum of the fractional deviations from targets or the sum of squares of those deviations all represent surrogates for minimizing losses.

We chose initially to minimize the maximum fractional deviation or the maximum deviation. Now we show how to minimize the sum of deviations, the sum of deviations squared, the sum of fractional deviations, and the sum of fractional deviations squared. We show all these forms because no form can be so firmly defended or recommended to justify describing it and not others. We develop the deviation from the target as the slack variable whose value separates the irrigation release from the target; that is, where u_k is the shortfall from the irrigation target, t_k, in month k. The mathematical equation defining u_k is

$$z_k + u_k = t_k \qquad k = 1, 2, \ldots, 12 \qquad (5\text{-}10)$$

The minimization of the slack variables or deviations is

$$\text{Minimize} \qquad z = \sum_{k=1}^{12} u_k \qquad (5\text{-}11)$$

And the minimization of squared slack deviations is

$$\text{Minimize} \qquad z = \sum_{k=1}^{12} u_k^2 \qquad (5\text{-}12)$$

Of course, piecewise approximation of the objective that has the squared terms would be needed. Each objective would be minimized, subject to all of the constraints defining the problem as well as to the definitional constraint for the slack deviations, (5-10) earlier. The objective of minimizing squared

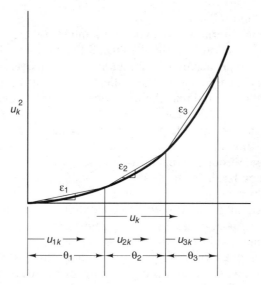

FIGURE 5-1   Piecewise approximation of u_k^2.

slack deviations is possible because the square of u_k is convex and can be piecewise approximated (Figure 5-1). From this figure, in which all k variables are approximated by the same length divisions:

θ_j = length of segment j
ε_j = slope of segment j
u_{jk} = portion of segment j, shortfall variable k, that is filled

The restructured minimization of squared slack deviations is

$$\text{Minimize} \qquad z = \sum_{k=1}^{12} \sum_{j=1}^{3} \varepsilon_j u_{jk} \qquad\qquad (5\text{-}13)$$

$$\text{s.t.} \qquad u_k - \sum_{j=1}^{3} u_{jk} = 0 \qquad k = 1, 2, \ldots, 12 \qquad (5\text{-}14)$$

$$u_{jk} \le \theta_j \qquad k = 1, 2, \ldots, 12; j = 1, 2, 3 \qquad (5\text{-}15)$$

and this objective is optimized subject to all the constraints described earlier.

To minimize the sum of fractional deviations and of fractional deviations squared, we need to minimize either

$$z = \sum_{k=1}^{12} u_k/t_k \qquad (5\text{-}16)$$

or

$$z = \sum_{k=1}^{12} (u_k/t_k)^2 \qquad (5\text{-}17)$$

The former objective needs no further work; it is already linear. The later objective is convex, as needed for minimization and should be piecewise linearized.

Before leaving the topic of shortfalls from the irrigation target, we can now comment on our earlier observation that shortfalls in June could lead to decreased needs in July. All of these irrigation shortfall objectives assumed the targets were known in advance. The targets could, however, be functions of previous releases, shortfalls or fraction shortfalls. For instance, the target in k might be diminished in a linear fashion by a function of the shortfall in $k - 1$. That is,

$$t_k = t_k^o - \delta u_{k-1} \qquad (5\text{-}18)$$

where t_k^o is the irrigation target in k if there is no shortfall in the previous month or interval. The appropriate modification for shortages in the previous two periods might also be linearly approximated but the joint or product effects of the two successive previous shortages seems more reasonable.

The reader should note that this treatment of the shortfalls from an irrigation target is quite general and could be applied to other reservoir functions besides irrigation. If shortfalls from the streamflow requirement, f_k, were permissible, the various methods could also be used to model these shortages in the objective function.

THE RECREATION OBJECTIVE

The recreation objective will be served if fluctuations in storage and thus in the height of the shoreline can be kept in as narrow a range as possible. The notion of a narrow range can be translated in a number of ways. In considering the objective involving storage or height of the water surface, we assume it is sufficient to operate on the end-of-month storages or end-of-month heights. If it is not, the monthly average of prior and current end-of-month storages or of prior and current end-of-month heights can be used in the constraints and objectives we describe.

We will develop objectives that could apply to changes in storage, but they could apply as well to changes in water surface heights. One objective might be to minimize the sum of the absolute values of month-to-month changes in storage. Another objective might be to minimize the sum of squares of these same variables. A third (and related) objective would be to minimize the maximum of the month-to-month changes in storage. A fourth objective might be to minimize the sum of deviations of storages from the mean storage or from a preset target for recreational storage. The last objective will be to minimize the range over the record of storage values.

To represent the water surface height, we use the same linear approximation of height as a function of storage that we used in the hydropower discussions of Chapter 3, namely,

$$h = h_o + ms$$

Even though this is the height above the turbines, to translate the height to another datum requires only the addition of another constant term. Thus, operating on this approximation of water surface height will give the same answer to objectives involving changes in height or deviations of height.

Rather than structure the objectives both for storage and for water surface height, we observe that all of the constraints needed to define the objectives for storage can be written for water surface height if a set of definitional constraints

$$h_t = h_o + ms_t \qquad t = 1, 2, \ldots, n$$

are written for all heights and appended to the problem in standard form

$$h_t - ms_t = h_o \qquad t = 1, 2, \ldots, n \qquad (5\text{-}19)$$

Because the translation to height is straightforward and immediate, we proceed to describe the four objectives only for storage—understanding that the reader is aware that the extension to water surface heights requires only using constraints (5-19) to define the height of the water surface.

We first structure the objective of minimizing the sum of the absolute values of month-to-month changes in storage. For this objective, we need to define positive and negative changes in month-to-month storage; that is,

$$s_t - s_{t-1} = v_t^+ - v_t^- \qquad t = 1, 2, \ldots, n \qquad (5\text{-}20)$$

where the variables

v_t^+ = positive when $s_t > s_{t-1}$
v_t^- = positive when $s_t < s_{t-1}$

This constraint is written in standard form as

$$s_t - s_{t-1} - v_t^+ + v_t^- = 0 \qquad t = 1, 2, \ldots, n \qquad (5\text{-}21)$$

And the minimization of the absolute values of changes in storage is simply

$$\text{Minimize } z = \sum_{t=1}^{n} v_t^+ + \sum_{t=1}^{n} v_t^-$$

subject to the preceding constraints (5-21) and all of the preceding constraints (5-1) to (5-6) that define the requirements portion of the problem.

If it is desired to minimize the sum of squares of month-to-month changes in storage, that is,

$$\text{Minimize } z = \sum_{t=1}^{n} (v_t^+)^2 + \sum_{t=1}^{n} (v_t^-)^2 \qquad (5\text{-}23)$$

we would then first need to piecewise approximate the convex functions represented by $(v_t^+)^2$ and $(v_t^-)^2$, as we have described earlier in this chapter. And, of course, the other definitional constraints for v_t^+ and v_t^- and other required constraints would be included.

We can also minimize the maximum of the month-to-month changes in storage by use of the following constraints, which define the maximum along with the objective

$$\text{Minimize} \qquad z = y \qquad (5\text{-}24)$$
$$\text{s.t.} \qquad s_t - s_{t-1} \leq y \qquad t = 1, 2, \ldots, n \qquad (5\text{-}25)$$
$$s_{t-1} - s_t \leq y \qquad t = 1, 2, \ldots, n \qquad (5\text{-}26)$$

where y is the unknown maximum change in storage. Also we would need to include the constraints required to define the rest of the problem. Constraints (5-25) suggest that the largest of the *increases* in storage is less than or equal to y. Constraints (5-26) suggest that the largest of the *decreases* in storage is less than or equal to y. These two constraints together are needed to define the maximum month-to-month change.

We also can minimize the sum of deviations of storage from average storage or from some target storage for recreation. We let

g = average storage for recreational purposes over the record
or
g = target storage for recreation

If g is the average storage over the record, it is defined in standard form by

$$ng - \sum_{t=1}^{n} s_t = 0 \qquad (5\text{-}27)$$

If g is a prespecified target, we can proceed directly to minimizing the deviations from the target without constraint (5-27). This form utilizes definitions of the deviations, which are, in standard form,

$$s_t - r_t^+ + r_t^- = g \qquad t = 1, 2, \ldots, n \qquad (5\text{-}28)$$

where

$r_t^+ =$ deviation of storage above g (excess over g)
$r_t^- =$ deviation of storage below g (decrement below g)

The definition in standard form, if g is defined as the average storage according to (5-27) becomes

$$s_t - r_t^+ + r_t^- - g = 0 \qquad t = 1, 2, \ldots, n \qquad (5\text{-}29)$$

Given the definition of deviations (5-28) and (5-29), and, if needed, a definition of average storage (5-27), the objective of minimizing deviations from a target or average storage is

$$\text{Minimize } z = \sum_{t=1}^{n} r_t^+ + \sum_{t=1}^{n} r_t^- \qquad (5\text{-}30)$$

The minimization of the squares of these same variables simply requires piecewise approximation of their convex functions—with the appropriate definitional constraints for the piecewise approximation process.

The last objective we describe—although undoubtedly there are others—is to keep storage in the tightest possible range. We can achieve this by adding constraints to the problem that define the unknown upper bound and the unknown lower bound for storage. Thus, we write

$$s_t \leq s_U \qquad t = 1, 2, \ldots, n \qquad (5\text{-}31)$$

$$s_t \geq s_L \qquad t = 1, 2, \ldots, n \qquad (5\text{-}32)$$

where

$s_U =$ unknown upper bound on storage
$s_L =$ unknown lower bound on storage

The difference between s_U and s_L is to be minimized, that is,

$$\text{Minimize } z = s_U - s_L \qquad (5\text{-}33)$$

The integrated reservoir problem can be summarized as the optimization of these objectives subject to (5-1) to (5-6) and nonnegativity constraints. The trade-off between the objectives can be developed by the constraint method or by the weighing method.

As an example of two objectives that might be traded off against one another, we might choose one irrigation objective, such as minimizing the maximum fractional shortfall, over all months, and one storage/recreation objective, such as minimizing the range of storages over the years of operation.

Thus we would have

Minimize $\quad z_1 = U$ (the maximum fractional irrigation shortfall)

$\qquad\qquad z_2 = s_U - s_L$ (the range of storage values)

s.t. $\qquad t_k U + z_k \geq t_k \qquad k = 1, 2, \ldots, 12 \qquad\qquad (5\text{-}9)$

$\qquad\qquad s_t - s_U \leq 0 \qquad t = 1, 2, \ldots, n \qquad\qquad (5\text{-}31)$

$\qquad\qquad s_t - s_L \geq 0 \qquad t = 1, 2, \ldots, n \qquad\qquad (5\text{-}32)$

and (5-1) to (5-6). Application of the constraint method or weighting method is described in Cohon (1978) and Cohon and Rothley (1997).

A number of additional objectives might be considered. One objective might be represented by the capacity of the reservoir. If the reservoir were not yet built, if this were a hypothetical analysis of a system under consideration, or if we were considering raising a dam height, then we could consider the impact of allowing a larger or smaller reservoir. We could then observe how the reservoir capacity would influence the trade-off between the irrigation objective and the storage/recreation objective. Adding a third objective expands the trade-off possibilities to three in number: irrigation versus capacity, storage/recreation versus capacity, and irrigation versus storage/recreation. Nonetheless, the information is easy to collapse to a single graph. The single graph would have a vertical axis of maximum fractional shortfall from the irrigation target and a horizontal axis of the range of reservoir storages. The trade-off curves between these two objectives would then be provided for a number of values of reservoir capacity with each capacity producing a potentially different trade-off curve.

The solution developed, however, will not include a storage-rule curve, as we have structured it so far. To provide the rule curve, we rewrite constraints (5-1), (5-2), (5-3), and (5-5) with just 12 storage values as in the formulations in previous chapters.

BIBLIOGRAPHY

Cohon, J. 1978. *Multi-objective Programming and Planning.* Academic Press, New York.

Cohon, J., and K. Rothley. 1997. "Multi-objective Methods." In *Design and Operation of Civil and Environmental Engineering Systems,* C. ReVelle and A. McGarity, Eds., pp. 513–566. John Wiley, New York.

PART II

RESERVOIR ALLOCATION AND REALLOCATION
Including the New Model for Reservoir Reliability

In Part I the analysis has focused primarily on situations in which the reservoir or reservoirs were in place and the uses of that fixed-size reservoir were to be optimized. Early on, in the chapters on reservoirs directed primarily toward water supply operation, we did show how to find the smallest reservoir, or least cost set of reservoirs, that delivered a given monthly yield or a distributed annual yield. And a similar analysis was performed for the delivery of hydropower. Our analysis was conventional and noneconomic in the sense that we did not value the water yield or value the hydropower produced but conceptually traded these quantities individually (and in their fundamental units) against the capacity or the cost of the system.

There is another way to approach problems of deciding yields, flood storage volumes, hydropower production, and the like. It is a decidedly economic perspective that we provide next. Within this view, the reservoir becomes an enterprise, resembling a factory that has multiple products, a factory that is yet to be built and whose range and amounts of various products are yet to be determined. Alternatively, as we show late in this part, the factory may already be built with some machines of fixed size already in place. The factory may come under review for decisions on whether the same products in the same quantities should still be provided or whether some alteration in the product mix is appropriate—because prices may have changed.

In the terminology of government agencies that cause reservoirs to be built and decide on their product mix and levels of production, the problem of how large to build, what services to provide, and at what levels to provide them—this problem is called allocation. When the reservoir is in place and the decisions are only on possible modification of the levels of services to provide, the problem is called reallocation.

There are two problems, allocation and reallocation, to be taken up in this second part of the book. These turn out to be difficult problems when approached properly. The proper solution to these problems required the creation of a new and fundamental model for reservoir design/reservoir operation. This model, the need for which is motivated in the next chapter, occupies a considerable portion of this part of the book. Although the model is needed for the allocation process, the formulation and methodology is also a stand-alone model that can be used for basic water supply planning.

This is the plan of Part II. Its opening chapter, Chapter 6, examines allocation in a new way, using mathematical tools that existed until the present. A deterministic allocation can be achieved with existing tools, but the required reliability of the water supply service is exceedingly difficult to investigate and incorporate into the analysis. The method of synthetic hydrology to handle water supply reliability is shown to be extremely burdensome to apply effectively when flood control services are part of the production mix of the reservoir. This difficulty motivates the need for the creation of a new model for water supply operation/design in the next chapter.

In Chapter 7, we focus only on the water supply service and introduce the new methodology that constrains water supply reliability; the advantages of the new methodology are explained. The model in this chapter, as in the first chapter of the book, is a mathematical program, an optimization model, but in this chapter, the model is augmented by chance constraints. These chance constraints ensure that storages, using operation over the historical record, are greater than or equal to zero with a specified level of reliability. To write these constraints, conditional densities of streamflows are created by using the correlation structure of successive inflows and the immediately preceding historical flow(s). The properties and uses of the methodology in the context of operating or determining the capacity of a single reservoir serving only the water supply function are explored in some detail, since the model is quite general.

In Chapter 8, the new model is incorporated in the allocation process allowing proper treatment of water supply reliability. The difficulty of constraining water supply reliability in the context of the added function of flood control is seen to disappear. In Chapter 9, with allocation on a firmer footing, we are able to focus on reallocation of the reservoir's services. It is admittedly a somewhat tortuous path we are describing. Nonetheless, it is an honest rendition of the investigation process or, put another way, of how the question of how to do allocation and reallocation came to be at least partially answered.

CHAPTER 6

ALLOCATING RESERVOIR SERVICES AMONG WATER SUPPLY, FLOOD CONTROL, AND HYDROPOWER: THE DETERMINISTIC FLOW ENVIRONMENT

INTRODUCTION

In this chapter, we consider the allocation of reservoir services among water supply, flood control, and hydroelectric energy production. We assume no dam is yet in place, but that water is a valued commodity for some municipal or industrial user. We also assume that flood control benefits in the form of losses averted will derive from maintaining free volume in the reservoir. Finally, hydroelectric energy is valued in the marketplace. Although few new reservoirs are being built in the United States at the present time, the methodology we will develop will be seen as easily capable of adaptation to the process of reallocating reservoir services from an existing facility. Reallocation of reservoir services for an existing reservoir has been studied by Wurbs and Cabezas (1987). These authors sought to minimize the sum of economic losses due to water shortages and due to flooding in the context of hydrologic simulation of reservoir inputs. As well, the U.S. Army Corps of Engineers (1988) has identified situations in which flood pool volumes might be seasonally adjusted to derive greater economic benefits from existing reservoirs (see also Johnson et al., 1989).

The division of reservoir services among water supply, flood control, and hydropower production requires both a reservoir model and an economic objective. The first positive term of the objective is the value of water sold in a year. The second positive term is the revenue derived from the sale of hydroelectric energy. From these quantities is subtracted both the expected annual losses that are associated with flood pool volume as well as the annualized cost of the reservoir. In the first section of this chapter ("Allocation

Between Water Supply and Flood Control") when only water supply and flood control are considered, the entire reservoir volume may be conceptually divided into a volume available for water storage and a volume available to store floods. These two volumes can also be allocated as different values by month, but the sum of the two volumes always totals to the entire capacity. A reservoir model that optimally allocates storage(s) between water supply and flood control is then developed.

In the second main section of this chapter ("Allocation Extended to Water Supply, Flood Control, and Hydropower"), we add the hydropower function to the analysis and allocate reservoir services between all three functions: water supply, flood control, and hydropower. We do not, in this second case, attempt any conceptual division of the storage volume of the reservoir.

We develop both of these analyses for a to-be-built reservoir rather than an existing reservoir in order to think clearly about the process of allocation. Even though environmental impacts are an obvious by-product of building a new reservoir, we do not include these in the present formulation. It would not be hard, however, to append to the models being developed minimum streamflow requirements for maintenance of aquatic species. Other environmental impacts, such as habitat loss, would be more difficult to consider.

ALLOCATION BETWEEN WATER SUPPLY AND FLOOD CONTROL

The allocation requires a backbone of streamflows that will be taken to be the historical record, suitably modified to reflect development changes along the river of interest. In the initial model of this chapter, no constraint will be placed on the reliability of the flood control function (i.e., no constraint on the probability of spillway use). The reliability of the water supply function can be handled by the use of synthetic hydrologic sequences in the simplest case considered here. In all other cases, it appears that with the present mathematics of synthetic hydrology, the placement of a reliability constraint on water supply is not an option that can be exercised easily. The creation of that option, the possibility of imposing a water supply reliability constraint, is the purpose of Chapter 7.

In all of these models, the reader will have the first impression that the price that a municipal and industrial water user will have to pay for water is assumed unknown. This is a device used only to derive further information. Its use will be explained. For now, the specification of price only appears to be restrictive; it will turn out not to be.

Model 6-a: Flood Pool Volume Constant and Water Requirement Constant through the Year

The first model to be structured is the simplest. A steady supply requirement, month to month, is imposed, and the flood control volume is the same value for each month. Needed notation for the model are

c = capacity of the reservoir, the sum of water supply storage and free volume in units such as billion gallons or cubic kilometers (either given or unknown)

v = volume reserved for flood control in all months (unknown)

$f(v)$ = expected annual flood losses if the volume v is reserved for flood control in every month

$g(c)$ = amortized annual cost of a reservoir of volume c

p_w = price per billion gallons for water supply water, a parameter that will be varied

q = steady release in billions of gallons of water per month that can be economically justified at a price of p_w, unknown

i_t = historical monthly flows for the months $t = 1$ to n, inclusive, billion gallons

w_t = water released in month t in excess of q billion gallons in order to maintain water storage less than or equal to $c - v$

s_t = storage of water at the end of month t

s_o = unknown initial storage of water

In nonstandard form, the problem is

Maximize $z = 12p_w q - f(v) - g(c)$ (*annual* benefits)   (6-1)

s.t. $s_n \geq s_o$ (6-2)

$s_t = s_{t-1} + i_t - q - w_t$ $t = 1, 2, \ldots, n$ (6-3)

$s_t \leq c - v$ $t = 1, 2, \ldots, n$ (6-4)

We will assume for purposes of the present model that controlling the flood pool at the end of each month (6-4) is sufficient to maintain the pool throughout the month. We could also constrain average storage as well as end-of-period storage. Although these steps should come close to achieving the desired volumes, constraints could be added to assure proper maintenance of the flood pool throughout the month if data on shorter duration flows were available.

The objective consists of three terms. First is the annual revenue from the sale of q units of water each month. This revenue is simply the price times the quantity sold. The second term is the expected annual flood losses if a free volume of v is maintained throughout the year. The third term is the amortized annual cost of a reservoir whose total capacity (water storage plus free volume) is c. The second and third terms are, of course, subtracted from the first.

In the objective, the monthly quantity of water sold q is unknown, the free volume v is unknown, and capacity c is unknown. If the cost function $g(c)$ is of a shape incompatible with linear programming optimization, the problem could be solved iteratively for various values of c. If capacity c were known, there would be only two terms and the problem would be to decide the division of c between flood storage v and water supply storage $(c - v)$.

The first term of the objective is linear in q. That is, each additional unit of water sold is presumed to sell for the same price. The second term is likely to be convex in v; that is, losses decrease with v at a decreasing rate or increase with v at an increasing rate. Since we are maximizing the negative of a convex function, this is equivalent to maximizing a concave function. As pointed out earlier, the technique of linear programming can be utilized in such a situation by approximating the function with short linear segments, whose incremental variables will always enter in the proper order. The third term, after an initial charge, is likely to be convex, and so, once again, its negative is concave, and linear programming via piecewise approximation can be applied. Probably a lower bound constraint on c, placing the function into its convex portion is appropriate to append to the constraints.

The three types of constraints serve the following functions. The first constraint ensures that the ending reservoir storage is greater than or equal to an unknown initial storage, so that water is not "borrowed" from initial storage to increase benefits. The second constraints are the mass balance constraints that relate end-of-month storage to the beginning-of-month storage and any inputs or releases during the month. The last constraints limit water storage to no more than that portion of c that is *not* reserved for flood storage.

In standard form the problem may be restated as

$$\text{Maximize} \quad z = 12p_w q - f(v) - g(c) \quad \quad (6\text{-}1)$$

$$\text{s.t.} \quad s_n - s_o \geq 0 \quad \quad (6\text{-}5)$$

$$s_t - s_{t-1} + q + w_t = i_t \quad \quad t = 1, 2, \ldots, n \quad \quad (6\text{-}6)$$

$$s_t - c + v \leq 0 \quad \quad t = 1, 2, \ldots, n \quad \quad (6\text{-}7)$$

This model will allocate reservoir services (as well as storage in this special case) between water supply q and free volume v and determine total capacity c if the price of water is known, the loss function is known, and the reservoir cost function is known. The loss function and the cost function are, in fact, likely to be known. Less likely to be known is the price a customer will pay for water. This last circumstance leads us to the following kind of investigation.

Suppose we experiment with values of p_w. If we solve the problem at low values of p_w, our expectation is that not much water will be sold each month and the proportion of the reservoir devoted to flood storage will be relatively large. In general, as the price p_w is increased, the monthly quantity of water that would be projected to be sold for Municipal and Industrial (M&I) purposes would increase and the proportion of the total capacity devoted to flood storage should decline. Since the purchaser in fact, may not know what to offer for water, the iterative specification of p_w can be put to use to generate a supply curve. A supply curve is the response in terms of quantity available as a function of the price being offered for a good. Its general shape has the obvious form of more of a good being provided as the price offered increases. The supply curve for M&I water services might look like that in Figure 6-1, as derived from solving multiple linear programs at specified values of p_w. The curve shown is an engineer's view of a supply curve in which the independent variable, the price, is plotted on the abscissa, and the dependent variable, the amount of water that can be provided, is plotted on the ordinate. Economists traditionally reverse the axes.

What can be done with this supply curve? One way to think about the outcome of the analysis is to compare the amount of water, q_w, that would be made available at the price that the purchaser is willing to pay, p_w, to the amount the purchaser has requested q_R. If q_w exceeds q_R, the purchaser can obtain all the water it perceives is needed. If q_w is less than q_R, the purchaser will not be able to obtain the water in the quantity desired because the price the customer is offering is too low.

Suppose, on the other hand, that no price has been specified by the purchaser, just an interest in obtaining water. The purchaser might have asked: "What price would we have to pay for X million gallons per day (mgd) of water?" In a very real sense, the process of examining the division of reservoir services at different prices of water is an informative one to the potential purchaser since it reveals the value the purchaser would have to place on the water in order to compete successfully against the other services being obtained from the reservoir.

The situation could be reversed as well. Instead of a potential purchaser expressing interest in a water supply, the U.S. Army Corps of Engineers could initiate an allocation study and develop the curve of water availability versus price. Potential M&I water supply customers could then be approached with the information that X mgd could be available at price p_x or Y mgd at price p_Y. In economics terminology, this information is known simply as the supply curve.

An economist who felt that water demand was governed by price; that is, that water use was elastic, might suggest that the purchaser of M&I water

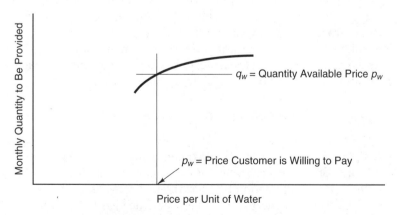

FIGURE 6-1   Supply curve for water.

would be willing to buy more as the price is reduced, that is, that the purchaser has a demand curve. Even though the demand curve is not necessarily revealed, the choice of quantity to be purchased, given the supply curve, represents the intersection of the demand curve with the supply curve.

What is being said here is that the development of the supply curve for M&I water services is the key set of steps in the process of allocation. It is the interaction of the M&I water customer with this curve that determines where on the supply curve the allocation will land. If it lands at q^*, the price is p^* and the allocation is the solution of the mathematical program with price $= p^*$. The solution specifies not only q^*, but also free volume v, capacity c, and water supply storage volume $(c - v)$ that one associates with price p^*.

Suppose now that the M&I water purchaser specifies a reliability with which water is needed. The model structured so far has used only the historical record. It would be possible to take the water supply volume $(c - v)$ and the requirement q and analyze the reliability of the combination using multiple synthetic sequences. (See Fiering, 1967; Fiering and Jackson, 1970.)

The process of allocation between flood control and water supply services with a constraint on water supply reliability turns out to be, in this simplest case, quite labor-intensive. Let us say that the steady release for water supply must be 90% reliable. We begin by generating 100 synthetic sequences, presuming them to be equally likely traces of flow and fully duplicative of the statistical properties of the parent or historic flow.

The repetitive process begins with a specification of a water supply storage volume m, which we use to replace the previous $(c - v)$. For this value of m, and for each sequence, the problem of establishing the largest steady release q is solved. The problem may be solved by search/trial or by linear programming or mass curve analysis (see Chapter 1 for details). This is repeated 100 times, once for each sequence. One hundred values of q are established in this way. They are then ranked from smallest to largest in value. The tenth value measured up from the bottom of the list can be delivered in approximately 90 out of 100 possible futures (the 90 that can support higher values of q). It is the 90% reliable steady release. For this value of m, tabulate $12p_wq$. Repeat the process for many values of m.

Now enumerate many values of v, the flood storage volume, and tabulate $f(v)$, the expected annual losses associated with this volume. For each value of m, try all values of v. For each of these alternatives, tabulate the value of

$$12p_wq - f(v) - g(m + v)$$

Choose the combination with the maximum value of the objective. Repeat for each value of m. For each value of m, tabulate the above local optimum. Choose the (m, v) combination that has the highest value.

This is the cumbersome process that seems needed to establish the best division between dedicated water supply volume and flood control volume when a constraint on water supply reliability is required. This process was needed for the simple case of two unchanging dedicated volumes for water supply and for flood control. We will see that the process becomes unwieldy when the volumes are not dedicated for the entire year but may be exchanged in different months. For now, we return to our allocation model and make it one step more complicated.

Model 6-b: Flood Pool Volume Constant and Water Requirement Varies by Month through the Year

As in Model 6-a, the historical record of inflows forms the backbone of the initial analysis without any measure/consideration of reliability.

The new notation is

q_k = unknown quantity of water required in month of the year k

Q_A = annual unknown water requirement = $\sum\limits_{k=1}^{12} q_k$

β_k = proportion of the annual water requirement needed in month k

$$q_k = \beta_k Q_A$$

The equation that defines q_k in terms of β_k and Q_A can be incorporated as a constraint in the model, so that the new version of the model is

Maximize	$z = p_w Q_A - f(v) - g(c)$	(6-8)
s.t.	$s_n \geq s_o$	(6-9)

$$s_t = s_{t-1} + i_t - q_k - w_t \qquad t = 1, 2, \ldots, n$$
$$k = t - 12[(t-1)/12] \qquad (6\text{-}10)$$

$$s_t \leq c - v \qquad t = 1, 2, \ldots, n \qquad (6\text{-}11)$$

$$q_k - \beta_k Q_A = 0 \qquad k = 1, 2, \ldots, 12 \qquad (6\text{-}12)$$

where $[u]$ is the integer part of u. As in Model 6-a, iteration over the price p_w creates the supply curve from which the water customer chooses a price and quantity. The incorporation of the water supply reliability requirement follows much the same process as described in Model 6-a and, again, is computationally burdensome. In the next model, we will see that the reliability requirement creates a most difficult challenge.

Model 6-c: Flood Pool Volume Differs by Month and Water Requirement Constant

This model is the same as Model 6-a except the flood pool is allowed to differ in volume from month to month. The notation is the same as in Model 6-a except that

v_k = volume of flood pool in month k
$f_k(v_k)$ = losses in month k as a function of the volume of the flood pool

A long historical record of inflows is initially assumed as in Models 6-a and 6-b. In nonstandard form, the problem is

$$
\text{Maximize} \quad z = 12 p_w q - \sum_{k=1}^{12} f_k(v_k) - g(c) \tag{6-13}
$$

$$
\text{s.t.} \quad s_n \geq s_o \tag{6-14}
$$

$$
s_t = s_{t-1} + i_t - q - w_t \qquad t = 1, 2, \ldots, n \tag{6-15}
$$

$$
\begin{aligned}
s_t \leq c - v_k \qquad & t = 1, 2, \ldots, n; \\
& k = t - 12[(t-1)/12]
\end{aligned} \tag{6-16}
$$

where $[u]$ is the integer part of u.

This model determines the optimal value of flood storage by month as well as the maximum quantity of water that can be delivered steadily from the conservation pool at the specified price. From another perspective, the model calculates the volume of the conservation pool by month at each specified price level. Or, still more accurately, the model determines the optimal division of the reservoir volume c between water storage capacity and flood pool capacity by month. Once again, we have the issue of maintaining storage *within the month* less than the appropriate conservation volume, but we have, in addition, the issue of the transition, month-to-month, between flood pool volumes. We know that the flood pool does not change from v_9 to v_{10} at midnight of September 30, so a gradual transition is needed.

Both of these issues seem resolvable by the same expedient—dividing each month into shorter nearly equal increments of, say, five 6-day periods for a 30-day month, or four 6-day and one 7-day period for 31-day months or four 7-day increments for February with the last increment of flow counted over 7 or 8 days as may be the case. All end-of-interval storages within a month, except the last, would be constrained less than the unknown conservation volume for that month. The last such storage, which is also the first in the next month, would be constrained less than the interpolated value of conservation volume between the two consecutive months. Inflows within the intervals would either have to be known or estimated under some assumption about the way the flow arrives in a given month.

This model, like the ones that preceded it, incorporates no explicit water supply reliability requirement as it uses the historical record of inflows. It differs from the preceding models in that it allows flood volume to vary by month, a property the fourth model shares as well. We will see that incorporation of water supply reliability poses a formidable challenge in those models that allow the flood pool to vary by month.

Model 6-d: Water Demand Is Assumed to Vary by Month and Flood Pool Differs by Month

Once again, the historical record of inflows anchors the initial analysis. Notation is the same as Model 6-c except that we return to the notation of Model 6-b for the water requirement variables

q_k = the unknown quantity of water required in month k

Q_A = the annual water requirement = $\sum\limits_{k=1}^{12} q_k$

β_k = proportion of the annual requirement needed in month k

Again, we have that

$$q_k = \beta_k Q_A \qquad k = 1, 2, \ldots, 12$$

as a constraint of the model or as a necessary substitution. The model is

$$\text{Maximize} \qquad z = p_w Q_A - \sum_{k=1}^{12} f_k(v_k) - g(c) \qquad (6\text{-}17)$$

s.t.

$$s_n \geq s_o \qquad (6\text{-}18)$$

$$s_t = s_{t-1} + i_t - q_k - w_t \qquad t = 1, 2, \ldots, n; \\ k = t - 12[(t-1)/12] \qquad (6\text{-}19)$$

$$s_t \leq c - v_k \qquad t = 1, 2, \ldots, n; \\ k = t - 12[(t-1)/12] \qquad (6\text{-}20)$$

$$q_k - \beta_k Q_A = 0 \qquad k = 1, 2, \ldots, 12 \qquad (6\text{-}21)$$

This fourth model, like the third, allows a variable flood pool by month. The incorporation of water supply reliability in these two models is the next question in sequence. It is worth noting that for a reservoir of stated capacity and specified flood storage volumes by month, that the calculation of either the maximum steady release or maximum annual supply is, by itself, not a terribly difficult problem (see Chapter 1). For example, assume c and all v_k values are known. Then the problem of maximum steady release is

Maximize q

s.t. $s_n \geq s_o$

              $s_t = s_{t-1} + i_t - q - w_t$ $t = 1, 2, \ldots, n$

              $s_t \leq c - v_k$ $t = 1, 2, \ldots n; k = t - 12[(t - 1)/12]$

What is happening here is simply that the maximum allowable water supply storage volume is varying by month.

This model will help us understand the root of the difficulty of attaching a reliability constraint to the water supply requirement. The discussion of Model 6-a, in which flood pool and water supply volume are constant, suggested an iterative approach using synthetic sequences to develop the appropriate division of storage volumes. Multiple trials of synthetic sequences were seen to be needed to establish a value of q that could be delivered with a stated reliability given a water supply volume m. Since we now have a variable flood pool, we no longer are able to conduct these trials for a single m. To conduct these trials, we need to begin with a capacity c. Then we specify a set of 12 values for the v_k and run multiple synthetic records through the water supply reservoir with its varying allowable water storage volumes. A steady release value is chosen that achieves the desired reliability (say 90%) given these water storage volumes.

The set of v_k establishes losses. The product $12p_wq$ establishes revenue. The cost of the reservoir is calculated. From these numbers, an objective value can be calculated. It is not the best objective value that can be obtained, as it is dependent on the values of the v_k, which were arbitrarily specified. To obtain the best objective value requires a 12-dimensional search over the $12v_k$, where each trial involves multiple runs of synthetic sequences. And, of course, these results are still price-specific although repetitions of the search over v_k values need not be done.

We draw from this description the conclusion that—if water reliability is to be added to a reservoir model with varying flood pools—the process is extraordinarily burdensome using synthetic sequences and subject to much conceptual difficulty. The conclusion is that a new model of water supply reservoir operation or design is needed to accomplish the task effectively. In Chapter 7, we develop a model for a water supply reservoir that bypasses the need to develop and use synthetic sequences and makes it possible to solve allocation problems, that is, develop the supply curve, directly. This new model will offer an explicit constraint on reliability. In Chapter 8, we rebuild the models of this chapter with the new reservoir formulation of Chapter 7 that constrains water supply reliability explicitly and we show how the new supply curve is developed. Finally, Chapter 9 places the new model in the situation of calculating the reallocation of reservoir services for an existing reservoir.

For now, however, we extend the allocation process to include hydropower in addition to water supply and flood control. Again, however, we restrict the analysis to the deterministic flow environment.

ALLOCATION EXTENDED TO WATER SUPPLY, FLOOD CONTROL, AND HYDROPOWER

In the previous main section, we investigated the problem of reservoir sizing and the allocation of total volume between water supply storage and flood storage. In this section, we introduce the hydropower function to the reservoir allocation problem. Our assumption throughout this section will be that the water released through the turbines for power generation is not available for the water supply function, but that water for M&I use is drawn from the reservoir separate from the turbine intakes. We also begin with another assumption, one that will subsequently be relaxed, that only the firm hydro energy commitment is valued, that energy generated in excess of the monthly commitment, if it is sold, adds only a negligible amount to the revenue from the sale of hydro. This low additional revenue may be due both to a lower price for such energy and the limited amount generated in excess of the firm commitment. We choose to deal with a firm commitment rather than the sum of energy generated because contracts are generally let for firm energy delivery. Even though the monthly commitment is firm, the power may be produced only for peak daily demands.

Excess Energy Adds Negligible Value

The new variables needed are those from Chapter 3 on the hydropower function:

x_t = release through the turbines in month t
h_t = height of the water surface above the turbines at the end of month t

Our plan of attack will of necessity be a bit roundabout because the hydropower function is nonlinear; it is proportional to the product of average head and the water directed through the turbines, as discussed in Chapter 3. Because of the nonlinearity, for each price of water, we will have to search across a range of firm energy commitments to find that level of energy delivery that maximizes the value of the reservoir's services.

The following problem would be solved for a particular value of firm energy commitment, d, and a price of water for M&I use, p_w.

$$\text{Maximize} \quad z = 12 p_w q - f(v) - g(c) \tag{6-22}$$

$$\text{s.t.} \quad s_n \geq s_o \tag{6-23}$$

$$s_t = s_{t-1} - q - x_t - w_t + i_t \qquad t = 1, 2, \ldots, n \tag{6-24}$$

$$s_t - c + v \leq 0 \qquad t = 1, 2, \ldots, n \tag{6-25}$$

$$\alpha(h_o + (m/2)s_t + (m/2)s_{t-1})x_t \geq d \qquad t = 1, 2, \ldots, n \tag{6-26}$$

where

d = specified value of the minimal monthly energy production

h_o = intercept of the linear function approximating head as a function of storage, known

m = slope of the linear function approximating head as a function of storage, known

α = constant that includes the efficiency of the turbines and is equal to 2.73×10^{-3} multiplied by that efficiency, known

p_w= price M&I users are willing to pay for water, unknown

p_F = price at which firm hydroelectric energy from the reservoir is to be sold, known

The assumptions that p_w is unknown and that p_F is known reflects the general absence of markets for water and the presence of an active market for electricity. Constraint (6-23) prevents borrowing water from initial storage. Constraints (6-24) are the mass balance constraints. Constraints (6-25) restrict water storage to a value less than the difference between capacity and flood volume. Constraints (6-26) require a minimum level of hydropower energy to be delivered each month. Constraints (6-26) are nonlinear and require linearization for an optimal solution to be provided. The method for linearizing this product form is described in Chapter 3 and involves dividing each constraint (6-26) by x_t and manipulating to produce the form

$$d/x_t - (\alpha m/2)s_{t-1} - (\alpha m/2)s_t \leq \alpha h_o \qquad (6\text{-}27)$$

The first term (d/x_t) is recognized as convex, and since the entire left-hand side must be less than or equal to αh_o, a piecewise linearization of d/x_t will provide incremental variables that enter in the proper order for the solution to make sense. For details, see Chapter 3.

Logarithmic decomposition of the original product form is another option for dealing with the energy constraint. To use logarithmic decomposition requires us to use $\bar{s}_t$ in place of $(s_{t-1} + s_t)/2$ in constraints (6-26) and add constraints defining $\bar{s}_t$. That is, constraints (6-26) are replaced by two sets of constraints:

$$\alpha(h_o + m\bar{s}_t)x_t \geq d \qquad t = 1, 2, \ldots, n \qquad (6\text{-}28)$$

$$\bar{s}_t - (1/2)\, s_{t-1} - (1/2)s_t = 0 \qquad t = 1, 2, \ldots, n \qquad (6\text{-}29)$$

Constraints (6-28) are further modified by rewriting as

$$\alpha m\bar{s}_t x_t \geq d - \alpha h_o \qquad t = 1, 2, \ldots, n \qquad (6\text{-}30)$$

Logarithmic decomposition of the original product form of this constraint separates the left-hand side of constraint (6-30) into the sum of two concave functions. Since the sum of the concave functions is required to be greater than the right-hand side, the sense is one of maximization of the left-hand side. Hence, the piecewise variables used to approximate the two concave functions enter in the order of largest slopes first, which is to say, in the proper order to make sense. Details of logarithmic decomposition of the product form for the hydropower equation are given in Chapter 10 in ReVelle et al. (1995).

Whatever method is used to solve the problem defined by constraints (6-23) to (6-25) under the net revenue objective (6-22), the optimal solution provides four fundamental items: a reservoir capacity, a free volume, a steady delivery, and a net revenue. Already specified and known is the hydroelectric energy commitment, which we refer to as d_i because we will be repeating the solution for multiple values of d. Hence, the determined quantities are indicated as capacity c_i, free volume v_i, steady delivery q_i, and net revenue z_i. The net revenue, however, does not include revenue from the sale of electricity. At a price p_F for electric energy, this term is $p_F d_i$ per month, and its annual total value $(12 p_F d_i)$ should be added to the maximum net revenue z_i given d_i. The new net revenue is $z_i + 12 p_F d_i$.

The odds on this solution providing the optimal division of reservoir services among water supply, flood control, and hydropower are slim. It is more likely a suboptimal solution to this problem. To arrive at the optimal division requires an exploration of the objective values $(z_i + 12 p_F d_i)$ over a range of values of d_i. Thus, the problem would be solved many times for different values of d_i—for each value d_i, a new z_i value is determined and to it the electricity revenue $12 p_F d_i$ is added. A plot would then be constructed of $(z_i + 12 p_F d_i)$ versus d_i, and the maximum point would be selected with its value of d_i as the solution that divides reservoir services optimally among the provision of water supply, the delivery of hydroelectric energy, and the provision of free volume for flood storage. If the jth value of d_i, that is, d_j, gives a maximum revenue of z_j, then the optimal level of capacity is c_j, of free volume v_j, and of steady water supply delivery q_j.

Missing in this analysis is the value of "dump energy," the hydroelectric energy produced in excess of the firm energy and sold at a lower price. To account for the value of this energy, the analyst could proceed differently.

Excess Energy has Value

The dump energy for each month is calculated by subtracting d_j from the energy actually delivered. The energy actually delivered is the left-hand side of (6-26). This amount in excess of d_j, call it E_i for month i, is multiplied by p_E (E for excess over firm) and the quantities of dump energy and revenue from dump energy are totaled for the entire horizon. Since the dump energy differs month by month, we need to calculate an annual average value of the dump energy. We assume $(n/12)$ is an integer equal to the number of years of record. The sum of the E_i over all months of the record divided by the number of years is the average annual dump energy. This annual average dump energy is multiplied by p_E to give the average annual revenue from the sale of the dump energy. This annual average revenue from the sale of dump energy is added to revenue from the sale of firm hydroelectric energy. Call this revenue term $R_H(d_i)$ for hydrorevenue at the ith firm energy level d_i.

The combined objective $z_i + R_H(d_i)$ is calculated for each value of d_i and its maximum value determined, giving at once the optimal division of reservoir services and reservoir size.

Still, despite the complexity of this analysis, it is likely that an even better division may yet exist and may be determined by using ideas from Chapter 3 for iterative solution using anti-zigzagging and the specification that a storage rule curve be determined.

Excess Energy Valued and a Rule Curve Determined

In Chapter 3, we rewrote the hydropower constraint as an equality and firm power as an unknown. We did this by subtracting a surplus variable from the left-hand side and making the right-hand side, F, a variable to be determined. Thus,

$$\alpha[h_o + (m/2)s_{t-1} + (m/2)s_t]x_t \geq F \qquad (6\text{-}31)$$

became

$$\alpha[h_o + (m/2)s_{t-1} + (m/2)s_t]x_t - E_t - F = 0 \qquad t = 1, 2, \ldots, n \quad (6\text{-}32)$$

where

E_t = unknown excess energy above F produced in month t
F = unknown firm energy delivery for all months

Again, using p_F as the contract price for firm energy and p_E as the contract price for excess (or "dump") energy, the value of the revenue over the n months of record from the sale of hydroelectric energy is

$$np_F F + p_E \sum_{t=1}^{n} E_t$$

The goal, and it is easy to lose sight of with all the detail of the formulation, is to divide reservoir functions in such a way that the total value of all the reservoir's services is a maximum. The services, once again, are water supply— for which a revenue stream is obtained, flood control—which yields benefits from damages prevented or reduced losses, and hydropower—for which a revenue stream is also obtained. In addition, the amortized reservoir cost must be subtracted from the revenue streams and benefits—in consistent units. We structure the problem with revenue streams and loss streams extended over the entire record.

The revenue from the sale of water over the horizon of the n months of the historical record is $np_w q$. The losses forgone due to the prevention of floods is $(n/12)f(v)$, where $n/12$ is the number of years of the historical record, presuming that the first and last month of the record are consecutive months. And the revenue from the sale of hydroelectric energy is

$$np_F F + p_E \sum_{t=1}^{n} E_t$$

The annual cost of the reservoir $g(c)$ is multiplied by the number of years of the historical record, so that the total of revenues is given by

$$z = np_w q - (n/12)f(v) + np_F F + p_E \sum_{t=1}^{n} E_t - (n/12)g(c) \qquad (6\text{-}33)$$

Of course, a discount factor could be used in (6-33).

This function is to be maximized subject to the following constraints, which are not in standard form

$$s_t = s_{t-1} - x_t - q - w_t + i_t \qquad t = 1, 2, \ldots, n \qquad (6\text{-}34)$$

$$s_t \leq c - v \qquad t = 1, 2, \ldots, n \qquad (6\text{-}35)$$

$$s_n \geq s_o \qquad (6\text{-}36)$$

$$\alpha[h_o + (m/2)s_{t-1} + (m/2)s_t]x_t - E_t \geq F \qquad t = 1, 2, \ldots, n \qquad (6\text{-}37)$$

As we discussed in the hydropower chapter, Chapter 3, no matter how we choose to solve this problem, even a globally optimal solution would not be implementable because the historical record on which it is based will never be seen again. Because our goal is an implementable strategy, we once more invoke the idea of finding a storage-rule curve for the reservoir operator to follow through the year. The rule curve, as we indicated earlier, is a set of end-of-month storages for each of the 12 months of the year.

The problem can and should be restated then with only 12 storage variables:

$$\text{Maximize } z = np_w q - (n/12)f(v) + np_F F + p_E \sum_{t=1}^{n} E_t$$

$$- (n/12)g(c) \qquad (6\text{-}38)$$

$$\text{s.t.} \qquad s_k = s_{k-1} - q - x_t - w_t + i_t \qquad t = 1, 2, \ldots, n;$$
$$k = t - 12[(t - 1)/12] \qquad (6\text{-}39)$$

$$s_k \leq c - v \qquad k = 1, 2, \ldots, 12 \qquad (6\text{-}40)$$

$$s_{12} = s_o \qquad (6\text{-}41)$$

$$\alpha(h_o + (m/2)s_{k-1} + (m/2)s_k]x_t - E_t \geq F \qquad t = 1, 2, \ldots, n;$$
$$k = t - 12[(t - 1)/12] \qquad (6\text{-}42)$$

The 12 s_k values are estimated initially for use in the power equation, which, on insertion of the values, then becomes linear in x_t, E_t, and F. The linear program is solved and 12 new values of the s_k are returned for reuse in the power equation. The LP is solved repeatedly for each set of values of the s_k that are calculated. The process continues until the s_k values either stabilize or are found to "bouncing around." If zig-zagging does occur, the changes in values of the s_k can be diminished using the Equation (3-25) as the damping mechanism. By using (3-25), both the magnitude of changes and the number of steps to achieve stable values of the s_k should be decreased. Experimentation with the damping rule may be necessary so that rate of convergence toward stability is "not too fast and not too slow."

When 12 stable s_k have been obtained, the solution is at least a locally optimal division of reservoir services between water supply, with its steady delivery of q^*, hydropower, with its firm commitment of F^*, and flood control, with its free volume of v^*.

As we did in the first section of the chapter, this optimization has been accomplished using only the historical record of inflows. Thus, no measure of water supply reliability has been incorporated into this analysis that considers water supply, flood control, and hydropower. As we pointed out in the first main section, the mechanism to assess reliability is very cumbersome and becomes exceedingly difficult when flood storage pools vary by month. Once again, this difficulty of incorporating reliability motivates the development of an entirely new model of reservoir reliability, and this task is taken up in Chapter 7. In Chapter 8, the new model for reservoir reliability is incorporated into the allocation model of the present chapter. In the concluding chapter, Chapter 9, reallocation is studied using many of the ideas developed in Chapters 6, 7, and 8.

BIBLIOGRAPHY

Fiering, M., 1967. *Synthetic Hydrology.* Harvard University Press, Cambridge, Mass.

Fiering, M., and B. Jackson. 1971. *Synthetic Streamflows,* Water Resources Monograph No. 1. American Geophysical Union, Washington, D.C.

Johnson, W., R. Wurbs, and J. Beegle. 1990. "Opportunities for Reservoir-Storage Reallocation." *Journal of Water Resources Planning and Management (ASCE),* Vol. 116, No. 4, pp. 550–556.

ReVelle, C., E. Whitlatch, and J. Wright. 1995. *Civil and Environmental Systems Engineering*. Prentice Hall, Englewood Cliffs, N.J.

U.S. Army Corps of Engineers. 1988. *Opportunities for Reservoir Storage Reallocation*, IWR Policy Study 88-PS-2. Water Resources Support Center, Institute for Water Resources, Ft. Belvoir, Va.

Wurbs, R., and L. Cabezas. 1987. "Analysis of Reservoir Storage Reallocations." *Journal of Hydrology*, Vol. 92, pp. 77–95.

CHAPTER 7

ACHIEVING A RELIABLE WATER SUPPLY WITH CHANCE-CONSTRAINED PROGRAMMING AND EVENT RESPONSIVE DENSITY ADJUSTMENT

INTRODUCTION

An enduring problem in the management of water resources is the determination of the minimum reservoir capacity needed to deliver reliably a stated quantity of water. The problem has a parallel statement that seeks the largest yield that can be sustained with a reservoir of a given volume. From the time of Rippl (1883), the dominant mode of analysis has been to utilize the historical record of streamflows to determine the required capacity, given a specified yield. Fiering (1964) introduced synthetic hydrology, a methodology to generate flow sequences that are statistical look-alikes of the historical record. When a reservoir of a given capacity and stated yield was evaluated using as inputs a number of these equally likely records, the reliability of being able to deliver that yield could be estimated. Synthetic hydrology was quickly accepted as the methodology of choice for the evaluation of the reliability of a water supply reservoir or systems of reservoirs.

Of course, synthetic hydrologic records do not by themselves determine the reliability of a given yield. The stream flow, synthetic or historic, must be routed through the reservoir, producing the required yield and wasting excess inputs (the inputs that cannot be stored in the volume available) to downstream flow. If the routing of inflow results in any negative storage, the reservoir with its stated capacity is deemed unable to deliver the specified quantity of water and "failure" is declared. A count of the number of sequences that do not lead to failure can be converted to an estimate of the reliability of the yield.

The process of routing inflows through the reservoir to assess the maximum feasible water delivery schedule within the context of a given record, synthetic or historic, is called simulation. As an example, given the historical monthly inflows and a reservoir capacity, a specified feasible constant release (or release schedule by month) is drawn from the reservoir. Initial reservoir storage may be specified as a small volume of the reservoir if operation begins in a period of abundant stream flow. Given that no deviation is allowed from the constant release or from the release schedule, the trace of resultant reservoir storages is generated. If over the entire record, no storage less than zero or some other minimum level occurs, the release or release schedule is inflated incrementally. The process is then repeated to see if a negative storage level occurs. The process of operation and incremental inflation continues to be repeated until some storage in the record just hits zero. The constant release or release schedule is then at its apex. Further inflation would lead to negative storages.

However, there is a condition in which the constant release or release schedule could still be inflated. The condition obtains when occasional intervals of release less than the specified schedule can be tolerated. If violations of the release specifications that occur are counted through the length of the record, we have another crude measure of reliability—albeit one that depends on the allowed extent of violations of release requirements. Different allowable extents of violation will result in different levels of "reliability" (Moy et al., 1986). Reliability of the delivery of a release or of a release schedule is at the center of modern inquiry in reservoir management and the focus of this chapter.

In this chapter, the reliable yield of a water supply reservoir of stated capacity is newly investigated, recognizing the correlation structure of streamflow inputs to the reservoir. The methodology utilizes conditional densities derived from a combination of the correlation structure of the flows and the flows themselves as taken from the immediately preceding portions of the historical record. These conditional or adjusted densities are used in the formulation of a mathematical program constrained by chance whose backbone flows are the historical record. The yield is maximized subject to a constraint that storages are greater than or equal to some base level with a specified reliability. Example problems are solved to illustrate aspects of the methodology. The methodology applies a fairly standard set of predictive regression equations from streamflow modeling but can utilize any model that predicts future flow from past flow(s). The new mathematical programming formulation should find use in models of multipurpose reservoirs and in investigations of systems of reservoirs where a constraint on reliability is required.

The new form of chance-constrained programming developed here will be used to determine either (1) the smallest reservoir capacity needed to deliver a steady release or release schedule or (2) the maximum possible steady release or monthly release schedule that can be achieved from a reservoir of stated volume. This release or schedule will be delivered with an explicit constraint on the probability that each month's ending storage does not become negative. The technique, which should find general application in areas other than water resources, makes simultaneous use of both the explicit historical record in its entirety as well as the statistical parameters that can be derived from the historical flows.

Chance-constrained programming originated in the classic work of Charnes and Cooper (1959) in which stocks of heating oil were built in response to weather-dependent and hence random demands. Stocks were purchased according to a linear decision rule, a concept that was extended to water resources management by ReVelle et al. (1969) and ReVelle and Kirby (1970). The decision rule made release a linear function of current reservoir storage and/or inflows. Further development and application of the rule were undertaken by ReVelle and Gundelach (1975), as well as Houck et al. (1980), among others. The assumption in these models was that the reservoir(s) was (were) being used simultaneously for two functions that were in conflict—water supply and flood control. In contrast, in this chapter, the reservoir serves only the single function of water supply. Hence, release is logically treated as a single number or as a seasonal release schedule that is independent of current conditions of storage or flow. That is, release is not conditioned on the current status of the reservoir.

A related work by Hirsch (1978) utilizes two simulation methods to characterize reliabilities of adequate storage. In the first, the generalized risk analysis method, the probabilities of water emergencies, of use restrictions, or of water purchases are estimated using a reconstructed historical record that is routed through a reservoir of known capacity. In the second approach, known as position analysis, multiple flow records, which begin at the moment of concern, are used to assess the risks over the next few months that are associated with operation given the current status of reservoir storage.

PROBLEM SETTINGS FOR THE RELIABILITY CONSTRAINED MODELS

We begin the description by formally stating four fundamental problems that we will then structure in sequence. The problems are conceptually approached with the assumption that the reservoir (or reservoirs) are already in place at stated sizes, but, at least for the first two problems, it is easy to reverse the problem statement and assume that the release is given and the reservoir size is to be determined. Thus, Problem I is divided into two problems, Ia and Ib, and Problem II is stated in two forms, IIa and IIb.

Problem Ia: *Maximizing the Firm Release from a Single Reservoir.*
                 Given a long historical monthly inflow sequence and a capacity c, determine the maximum firm release q that can be achieved while maintaining the storage nonnegative with reliability level α.

Problem Ib: *Minimizing Reservoir Capacity Needed to Deliver a Firm Yield.* Given the same long sequence of historical flows and a steadily required flow q, determine the smallest reservoir capacity that will provide the flow and maintain storage nonnegative with probability α.

Problem IIa: *Maximizing the Annual Yield from a Single Reservoir with Water Requirements by Month.* Given the preceding information and a set of relationships defining relative water requirements by month, determine the maximum annual yield that can be achieved while maintaining storage greater than or equal to zero with α reliability.

Problem IIb: *Minimizing the Reservoir Capacity Needed to Deliver a Seasonally Proportioned Annual Yield.* Given stream flow data and fractional water requirements by month, and a stated annual yield, find the smallest reservoir that maintains storage nonnegative with α reliability.

Problem III: *Maximizing the Firm Joint Yield from Parallel Reservoirs.* Given historical flows and capacity information for a number of reservoirs, determine the maximum steady joint yield that can be achieved with probabilistic storage requirements for each reservoir.

Problem IV: *Maximizing the Joint Annual Yield from Parallel Reservoirs with Water Requirements by Month.* Given historical flows and capacity information for the reservoir system as well as a set of relationships defining the fraction of the annual water requirement to be delivered each month, determine the maximum annual yield that can be achieved with fractional allocation by month and probabilistic storage requirements for each reservoir.

Of these four problems only Ia, IIa, III, and IV will be formally structured. Problems Ib and IIb, will be described verbally, however.

Problem Ia: Maximizing the Firm Release from a Single Reservoir

To investigate Problem I, we let

s_t = storage at the end of period t, an unknown realization based on inflows and releases

S_t = storage at the end of t, a random variable

s_o = unknown initial storage level to which the reservoir will refill by the end of the record

i_t = inflow in month t, known from the historical record

w_t = water wasted to spill during period t, an unknown realization based on inflows and releases

q = release to water supply during each month, unknown

c = reservoir capacity, known

I_t = inflow in month t, unknown, a random variable whose conditional probability distribution function is known

Note that the inflow in month t appears twice in the listings of parameters, once as a random variable characterized by a probability distribution function, and once as the realization of that random variable as it occurred in the historical record. The storage at the end of month t also appears in a dual role, once as the realization associated with a given set of streamflows, capacity, and a decision on the steady release, and once as a random variable whose distribution depends on the release and the distribution of the immediately prior random streamflow. The distinction in the model between the historical realization of inflow and the random variable inflow is an important one, as is the distinction between realized storage and random variable storage.

If we were to seek the maximum feasible release to water supply over the record without knowledge of the randomness of inflow, the release would be predicated only on the historical flows as they actually occurred. The steady release would then be determined with perfect knowledge of the future (or the past, depending on your point of view). The value of the steady release would be an overly optimistic value because of the assumption that the future is known with certainty (or that the past will recur).

Of course, the problem can be viewed in another way. Given historical flow records, the maximum steady release could be specified in advance and the smallest reservoir capacity needed to sustain this release would be determined. This problem has been solved by the procedure known as "sequent peak" (Thomas and Burden, 1963) as well as by linear programming. The point, however, is that both of these methods rely on a record of flows that are either the historical flows or on a record that is generated by the set of methods collectively known as synthetic hydrology (Fiering, 1967).

In fact, at any moment in time, when release decisions must be made, all that is really known is the past. The decision on how much water can be made available for this month without damaging the reservoir's ability to respond later must be made at the start of the month despite the fact that the inflow in the upcoming month is not known. What is known, however, or rather what can be known if appropriate calculations are made, are the parameters of the distribution function (or density function) of the random inflow in the upcoming month (or some longer interval). In a real operating situation, the decision on release ought to be made on the basis of the parameters of this random variable. This month's release decision would not be/ought not be made solely on the basis of any particular past historical realization, but should be made based on both the realization and the statistical parameters relating future to past flows.

We structure Problem Ia in nonstandard form with a chance constraint as follows

(I)   Maximize $z = q$

$$S_t \leq c \qquad t = 1, 2, \ldots, n \tag{7-1}$$

$$S_t = S_{t-1} - q + i_t - w_t \qquad t = 1, 2, \ldots, n \tag{7-2}$$

$$P[S_t \geq 0] \geq \alpha \qquad t = 1, 2, \ldots, n \tag{7-3}$$

$$S_n \geq S_o \tag{7-4}$$

$$S_t \geq 0 \qquad t = 1, 2, \ldots, n \tag{7-5}$$

$$q \geq 0 \tag{7-6}$$

$$w_t \geq 0 \qquad t = 1, 2, \ldots, n \tag{7-7}$$

Constraints (7-1) ensure, no matter what release strategy is chosen, that the realized storage through the historical record with the calculated set of releases never exceeds the capacity of the reservoir. The means to ensure this is the spill or wasted release term, w_t, which appears in constraints (7-2). Constraints (7-2) are the mass balance or continuity relations that connect consecutive realized storage values through the input and output events that occur in a month. Constraints (7-3) are the chance constraints that require the random variable storage S_t greater than or equal to zero with α reliability. Constraint (7-4) ensures that the unknown ending storage is greater than or equal to the unspecified initial storage; this is a condition that prevents the total of releases from exceeding the total inflow over the period of record. Constraints (7-5), (7-6), and (7-7) are the feasibility constraints, written separately to facilitate discussion.

Although constraints (7-3) suggest that storage be greater than or equal to zero with a defined reliability, the base that it must exceed could be larger. It may be that volume is maintained for release to the river for environmental flow by or for navigation. In such cases, the storage measured here is to be over and above the volume maintained for the other purposes.

Each of the constraints (7-3) can be rewritten with the end-of-period storage random variable S_t replaced by the previous storage realization s_{t-1} less any release q plus the random inflow, I_t, in period t; that is,

$$S_t \text{ is replaced by } (s_{t-1} - q + I_t)$$

so that constraints (7-3) become

$$P[s_{t-1} - q + I_t \geq 0] \geq \alpha \qquad t = 1, 2, \ldots, n$$

or in the distribution function form

$$P[I_t \leq q - s_{t-1}] \leq 1 - \alpha$$

$$F_{I_t}(q - s_{t-1}) \leq 1 - \alpha$$

The linear deterministic equivalent is obtained by inversion and re-arrangement:

$$q - s_{t-1} \le F_{I_t}^{-1} (1 - \alpha)$$

where $F_{I_t}^{-1}(\mu)$ is the inverse of the distribution function evaluated at μ. The value $F_{I_t}^{-1} (1 - \alpha)$ will hence forth be replaced by $i_t^{1-\alpha}$, the flow value in month t exceeded with α probability or the $(1 - \alpha)$-percentile flow. The constraint is now written as

$$q - s_{t-1} \le i_t^{1-\alpha}$$

or, if q is a specified value, the constraint is

$$s_{t-1} \ge q - i_t^{1-\alpha}$$

If the level of reliability with which storage exceeds 0 is specified as 90%, the inverted distribution function can be replaced by $i_t^{0.10}$ or the value of inflow that is exceeded with 90% reliability. The linear deterministic equivalent can then be rewritten as

$$q - s_{t-1} \le i_t^{0.10} \qquad t = 1, 2, \ldots, n \qquad \text{(7-8)}$$

which replaces constraints (7-3) in the formulation of Problem I. The variable w_t is explicitly excluded from the representation of S_t in the chance constraint since during the critical record spill will not occur. In fact, its inclusion in the probabilistic storage constraint can lead to incorrect answers.

Problem Ia may now be written in standard form as

Problem Ia

$$
\begin{aligned}
\text{Maximize} \quad & z = q \\
\text{s.t.} \quad & s_t \le c & t = 1, 2, \ldots, n & \qquad \text{(7-1)} \\
& s_t - s_{t-1} + q + w_t = i_t & t = 1, 2, \ldots, n & \qquad \text{(7-2)} \\
& q - s_{t-1} \le i_t^{0.10} & t = 1, 2, \ldots, n & \qquad \text{(7-8)} \\
& s_n - s_o \ge 0 & & \qquad \text{(7-4)} \\
& s_t \ge 0 & t = 1, 2, \ldots, n & \qquad \text{(7-5)} \\
& q \ge 0 & & \qquad \text{(7-6)} \\
& w_t \ge 0 & t = 1, 2, \ldots, n & \qquad \text{(7-7)}
\end{aligned}
$$

Of course, this formulation can be restructured to minimize an unknown needed capacity c given a draft q that is to be delivered each month from the reservoir. This is the problem stated as Ib. Trading off capacity against draft provides the classical storage-yield curve, but in this case, in a probabilistic setting.

Constraints (7-8) and (7-5) serve related but different functions. The feasibility constraints (7-5) require that each end-of-period storage be greater than or equal to zero within the context of the chosen level of release and the historically realized values of inflow in the periods up to and including that period. Constraints (7-8) serve a parallel function relative to constraints (7-5), in that they require that each end-of-period storage is greater than or equal to zero with probability α. That is, end-of-period storage exceeds zero with a stated reliability within the random environment that characterizes flow during each period of operation.

Constraints (7-8) are designed to ensure that the end-of-period storage exceeds zero with a desired level of reliability. To achieve this, they must treat the inflow in the period as a random variable with a known probability distribution function. The probability distribution function to be utilized is not the unconditional distribution, but will be conditioned on previous inflow(s) that occurred in the historical record. In constraints (7-8), this month's inflow is not a predetermined value from the historical record. Instead, the realization of this random variable inflow is determined by making use of the inverse of the distribution function. In some months, constraint (7-8) (the probabilistic constraint) may be tighter than constraint (7-5); this would occur if the historical flow that occurs exceeds the $(1 - \alpha)$-percentile level of flow. In other months, constraint (7-5) may dominate because the historical flow turns out to be less than the $(1 - \alpha)$-percentile flow. The issue of which constraint binds is taken up in more detail later.

The constraint that storage be nonnegative with reliability α is really a constraint that states that the probability that release will have to be cut back during a month is less than $(1 - \alpha)$. The flow might have to be cut back because insufficient water is available to distribute. The way such a cutback might occur is that at some moment in the month, the cumulative value of the steady release in the month reaches the value of beginning monthly storage plus cumulative inflow up to that moment. From that moment on in the month, unless a large surge of flow occurs, further releases can be no more than current inflows, since negative storage values could obviously never occur. Thus, the probabilistic storage constraint is really a constraint on the probability of being able to deliver promised water through the period.

The need to utilize conditional distributions arises because, in the case of streamflow, the streamflow value in any month is known to be correlated with streamflow(s) in the month (or months) preceding the flow in question. Thus, in month t, the density and distribution function of the random variable flow can be characterized by the preceding streamflow(s) that actually occurred in the historical record.

For purposes of illustration in this discussion, we choose to assume that the random flow in month t can be best characterized by the flow in only the single preceding period. Although such a relation as we suggest is entirely reasonable, each river or stream would, in fact, have to be investigated to determine the best predictive correlation form. Prediction of this month's flow characteristics based on last month's realized flow serves to illustrate quite adequately the idea of a flow with known conditional distribution function. Prediction based on historically realized flows in the more distant past will not really add anything to our illustration of the concept, although in some particular situation, the use of additional flows more distantly past may be most appropriate.

To illustrate the derivation of the predictive form, we need to define the following typical pairs of observations of streamflows:

y_i = observed flow in February, year i
x_i = observed flow in January, year i

If there are 100 years of monthly streamflow records, then for each sequential set of months, there are 100 pairs of data. That is, there are 100 February–January pairs, 100 March–February pairs, and so on, except that there are only 99 January–December pairs. A linear regression is performed for each pair of months establishing the expected value of flow in month t given an observed flow in the previous month. For simplicity of presentation at this point in the discussion, we assume that streamflows are normally distributed.[1]

The flow that actually occurs is most likely not the expected value, but a flow larger or smaller than the expected value. The flow in a sequential pair of months can be modeled as

$$Y = aX + b + R$$

where Y is the predicted flow in the current month, and X is the observed flow in the month that preceded the current month. The terms a and b are the parameters established by the least squares regression, and R is a random shock, either positive or negative in value, that shifts the flow Y away from its expected value. This lag-one model is well-established in the literature of hydrology. See, for instance, Fiering (1967) or Fiering and Jackson (1971). The expected value of the random shock is assumed to be zero. The variance σ_R^2 of the random shock (which is derived in the Appendix) is estimated by

$$s_R^2 = s_Y^2 \left(1 - r^2\right)$$

---

[1] In fact, it is the logarithm of streamflow that is probably normally distributed because of the physical setting in which streamflow occurs.

where

s_R^2 = sample estimate of variance of the random shock
s_Y = sample standard deviation of the current month's flow Y
r = sample correlation coefficient relating the current and preceding
    month's flows

Thus, the random variable flow in the current month has the following parameters and is distributed with (expected value, variance) as

$$(ax + b, \; s_Y^2 (1 - r^2))$$

where x is the historical flow that occurred in the preceding month. To calculate the flow that is exceeded with 90% reliability, namely, $i^{0.10}$, we refer to the $n(\mu, \sigma)$ density. We know that the realization value with 90% of the area to the right is μ less 1.282 times the standard deviation of the random shock. Hence, the value of the flow in the current month with 90% of the area to the right is

$$i^{0.10} = (ax + b) - 1.282 s_Y (1 - r^2)^{1/2}$$

where x is the flow observed in the previous month, and $s_Y (1 - r^2)^{1/2}$ is the sample standard deviation. The method of estimating the conditional density from which the $(1 - \alpha)$-percentile flow is drawn is termed *event responsive density adjustment.*

If the logarithms of flows were normally distributed, the regression would be of the log of this month's flow against the log of last month's flow, creating a linear relation between the two. The expected value of this month's log flow would be based on the log of the flow realized last month. Similarly, a sample variance of the random shock of the log flow would be calculated. The 10-percentile log flow for the current month would be established in the same fashion as in the nonlog case, by subtracting from the expected value of the log flow 1.282 times the sample standard deviation of the random shock. The antilog of this number would then be used as the 10-percentile flow.

Problem IIa: Maximizing the Annual Yield from a Single Reservoir with Water Requirements by Month

This problem resembles Problem Ia in nearly all respects except that different proportions of annual yield are delivered in each month. In month k, the release is q_k, which is a known β_k proportion of the annual yield. That is,

$$q_k = \beta_k Q_A$$

where Q_A is the annual yield. Problem Ia can thus be directly modified with the addition of this definition of monthly yield; that is, Problem IIa is

Problem IIa

$$\text{Maximize} \quad z = Q_A$$

$$
\begin{array}{lll}
\text{s.t.} & s_t \leq c & t = 1, 2, \ldots, n & (7\text{-}9) \\
s_t - s_{t-1} + q_k + w_t = i_t & t = 1, 2, \ldots, n; k = t - 12[(t - 1)/12] & (7\text{-}10) \\
q_k - s_{t-1} \leq i_t^{0.10} & t = 1, 2, \ldots, n; & \\
& k = t - 12[(t - 1)/12] & (7\text{-}11) \\
q_k - \beta_k Q_A = 0 & k = 1, 2. \ldots, 12 & (7\text{-}12) \\
s_n - s_o \geq 0 & & (7\text{-}13) \\
s_t \geq 0 & t = 1, 2, \ldots, n & (7\text{-}14) \\
q_k \geq 0 & k = 1, 2, \ldots, 12 & (7\text{-}15) \\
w_t \geq 0 & t = 1, 2, \ldots, n & (7\text{-}16)
\end{array}
$$

where $[u]$ is the integer part of u.

All constraints have the same meaning as in Problem Ia except (7-12), which distributes the annual yield across the 12 months. The objective maximizes the annual yield. The "integer part of" notation is used to reduce monthly releases to just 12 values despite the fact that the release for each month is made $n/12$ times over the record of n total months. As before, constraints (7-11) require storage greater than or equal to zero with 90% reliability.

Problem IIb specifies Q_A and, hence, using Equation (7-12), the values of the 12 q_k. As a consequence, Equation (7-12) may be omitted. At the same time, capacity c is treated as an unknown value to be minimized.

Problem III: Maximizing the Firm Joint Yield from Parallel Reservoirs

The two previous models can be easily adapted to the operation in concert of multiple and parallel reservoirs where each is fully devoted to water supply. We illustrate the multiple-reservoir model with three reservoirs that are on roughly parallel and nonintersecting streams. Withdrawals are made from each reservoir and these withdrawals sum toward a common supply.

The new notation is largely parallel to the notation used for the single-reservoir model. We develop the model in two distinct stages as we did with Models I and II, beginning with the situation in which the demand for water is assumed constant through the year and proceeding next to the situation in which water demand varies through the year (as in Model II). In the first of these new models, we maximize the monthly yield. In the second of these models, we maximize the annual rather than the monthly yield. The notation for the model that seeks the maximum firm monthly yield from parallel reservoirs is

s_{jt} = storage in reservoir j at the end of period t, an unknown realization based on inflows and releases

S_{jt} = storage in reservoir j at end of t, a random variable

q_{jk} = release toward the joint water supply from reservoir j in month of the year k, unknown

i_{jt} = inflow to reservoir j in month t, known from the historical record

w_{jt} = water wasted to spill from reservoir j during period t, unknown

c_j = volumetric capacity of reservoir j, known

I_{jt} = inflow to reservoir j in month t, unknown, a random variable whose distribution, which is conditioned on previous inflow, is known based on a regression model

m = maximum over all months of the minimum joint yield.

With this extended notation, we can now structure a model that maximizes the amount of water that can be steadily supplied over the year from the three reservoirs. That is, we will seek the set of releases from each of the three reservoirs, specific by month of the year, that will provide the maximum over the year of the minimum of the monthly sum of releases. Notationally, we seek

$$m = Max. \left(Min_{k=1, 2, \ldots, 12} \left[\sum_{j=1}^{3} q_{jk} \right] \right)$$

Of course, we will operationalize this criteria. The mathematical program for this problem in nonstandard form with representational versions of the chance constraints is

Problem III

Maximize $z = m$

s.t. $s_{jt} \leq c_j$ $t = 1, 2, \ldots, n; j = 1, 2, 3$

$s_{jt} = s_{jt-1} - q_{jk} + i_{jt} - w_{jt}$ $t = 1, 2, \ldots, n; j = 1, 2, 3;$
$k = t - 12[(t - 1)/12]$

$P[S_{jt} \geq 0] \geq \alpha$ $t = 1, 2, \ldots n; j = 1, 2, 3$

$s_{jn} \geq s_{j0}$ $j = 1, 2, 3$

$$\sum_{j=1}^{3} q_{jk} \geq m \qquad k = 1, 2, \ldots, 12$$

$$s_{jt}, w_{jt} \geq 0 \qquad t = 1, 2, \ldots, n; j = 1, 2, 3$$

$$q_{jk} \geq 0 \qquad j = 1, 2, 3; k = 1, 2, \ldots, 12$$

$$m \geq 0$$

These constraints, except for the last structural constraint, parallel exactly those of Problem I. The last constraint defines the minimum monthly sum of releases. Of course, in standard form, m is shifted to the left-hand side of the equation. The chance constraints are operationalized in precisely the same way as in Problem I.

The solution of this problem, suitably altered and rearranged for application with a linear programming code, provides as its most critical output a release for each month of the year for each of the three reservoirs. In one month or season, one or two of the reservoirs may provide a greater share of the firm yield, and in another month or season, the dominant contribution(s) may shift. Nonetheless, the smallest sum of contributions/releases will be as large as possible.

In a larger sense, however, this problem is incomplete since the "system" can survive even if one reservoir's volume goes to zero. Still to be studied is the important problem of constraining system storage to be be greater than or equal to zero with α reliability. This latter problem appears to be challenging.

Problem IV: Maximizing the Joint Annual Yield from Parallel Reservoirs with Water Requirements by Month

This problem extends Problem III in the same way that Problem II extended Problem I. As in Problem II, different fractions of the annual yield are provided in each month. In month k, the sum of individual reservoir releases must be the β_k fraction of the annual yield. That is,

$$\sum_{j=1}^{3} q_{jk} = \beta_k Q_A$$

where Q_A is the unknown annual yield, as in Problem II.

This model looks precisely like Problem III except that the 12 constraints defining the relations of monthly to annual yield are appended; that is,

$$\sum_{j=1}^{3} q_{jk} - \beta_k Q_A = 0 \qquad k = 1, 2, \ldots, 12$$

and the objective is

$$\text{Maximize } z = Q_A$$

COMPARISON OF THIS APPROACH TO DESIGN
USING SYNTHETIC HYDROLOGY

The methodology proposed here offers potentially different answers to design questions than does the use of synthetic hydrology. The reasons for the potentially different answers reside in the differences between synthetic sequences and the α-reliable limits of inflows that are generated using historical realizations as the base of the projection.

A criticism that has been offered of synthetic sequences is the possible failure of such sequences to reproduce lengthy runs of low flows. Such persistent low flows in which successive monthly flows remain in their lower ranges may often be observed in historical records. These persistent low flows may, in fact, form the "critical period," the sequence of flows that defines and limits the maximum sustainable release. Nonetheless, if the persistence of low flows is not the dominant or defining statistical relation between flows, the generation scheme will not be based on that persistence and hence may fail to duplicate such long-term low flows.

This potential failure to reproduce persistent low flows may decrease the utility of synthetic sequences for design purposes. This is because the "critical period" may often be such a low flow sequence. The critical period determines the maximum release if the design is given, or it determines the design, if the steady release is given. With synthetic hydrology, multiple equally likely records are generated for purposes of probabilistic design or analysis. Suppose a design is determined based on the historical record (the smallest reservoir needed to deliver a stated yield through the historical record) and that multiple synthetic records are generated. For each record, a needed capacity for the reservoir is determined that delivers the stated consistent quantity of water. If the procedure for synthetic generation is not as likely to produce persistence as it should be, it will be necessary to generate more sequences to achieve the same depth of drought as exists in the historical record than would be necessary if a "correct procedure" were available. These additional sequences would potentially make the design to achieve a stated water supply seem to be more reliable than it really is. That is, if the design chosen delivers the stated consistent supply in 90 out of 100 synthetic sequences, the design is not necessarily 90% reliable. A "correct" generation procedure might have produced a sequence with drought depth this severe in 60 or 70 out of 100 synthetic sequences, making the real reliability of the design 60 or 70%.

The preceding argument is speculative since we do not have a methodology that produces guaranteed "correct" (statistically duplicative in every sense) sequences. The point is, however, that a failure to reproduce persistence in low flows may lead to designs that are not as reliable as they seem.

The design based on the historical record, however, is a design that reflects whatever persistence actually occurred in low flows. At the same time, though, the depth of the drought in the historical sequence is difficult to place in the hierarchy of possible drought depths, specifically because we do not have a methodology that guarantees sequences with low-flow persistence that mimics the low-flow persistence in the historical record. That is, synthetic hydrology will not guarantee us comparable sequences in terms of depth of drought. In part, this is a conundrum since what we desire is to reproduce extreme events comparable to the most extreme event that by definition is likely to have occurred only once.

A design based purely on the historical record of streamflows then has the distinct advantage of fully utilizing the most extreme low-flow sequence and the persistence that was actually displayed in the record. It follows that a probabilistic design methodology that uses the historical record as its base, that uses the low-flow persistence in the historical record in its calculation/ determination of the maximum possible steady release or the smallest needed reservoir capacity, will reflect more accurately the influence of persistence on that release or design. That is the case with the methodology proposed here.

The historic record might be thought of as a winding road of known curvature and position. We are concerned about possible probabilistic excursions that take place in a particular direction, that is, downward in flow. That is, the historical record stands as the base of the calculation, but potentially lower flows may occur. These lower flows are associated with particular levels of reliability of achieving nonnegative storage.

The probabilistic storage constraint

$$P[S_t \geq 0] \geq \alpha$$

is also written

$$P[s_{t-1} - q + I_t \geq 0] \geq \alpha$$

and has a deterministic equivalent of

$$q - s_{t-1} \leq i_t^{1-\alpha} \tag{*}$$

where $i_t^{1-\alpha}$ is the flow value in period t that is exceeded α fraction of the time. This deterministic equivalent is written in addition to the constraint implied by the deterministic nonnegative storage requirement (given operation over the historical record). The constraint implied by the deterministic nonnegative storage requirement is

$$s_{t-1} - q - w_t + i_t \geq 0$$

or

$$q + w_t - s_{t-1} \leq i_t \tag{**}$$

Most often no more than one of two constraints, (*) or (**) will be binding—if any storage constraints for month t bind at all.

Suppose the reliability level α is 0.90. This leads to use of a flow value that is exceeded 90% of the time; that is, the 10-percentile flow level. Since the 10-percentile flow is used in the deterministic equivalent, it should be roughly true that when one of these two constraints do bind, that only 10% of the time will the constraint that uses the historically realized flow be the binding constraint since on only 10% of the occasions should the historical flow be less that the 10-percentile value. The other 90% of the time, the binding constraint will be the deterministic equivalent constraint that uses the 10-percentile flow. This obtains because the 10-percentile flow is exceeded by the realized flow 90% of the time. Hence, the deterministic equivalent constraint, which uses the 10-percentile flow, is most often the tighter constraint (see Figure 7-1).

It seems useful to consider a picture of the historical flows and 10-percentile flows as they might occur through a portion of a hypothetical record (see Figure 7-2). If binding were to occur in period 4, the storage constraint using the historical flow would bind. If binding were to occur in period 11, the deterministic equivalent storage constraint would bind. This does not suggest that the two kinds of nonnegative storage constraints can be replaced by just one constraint for each month, that is, by the most binding constraint. Both kinds of constraints are generally still necessary. The constraint that utilizes the historically recorded flow, though it will be tighter only 10% of the time, is still needed in all cases to define the end-of-period storage in period t that is finally realized and that is used as the jumping-off point for defining the storage in the subsequent period. On the other hand, if the 10-percentile flow exceeds the historically realized flow, the deterministic equivalent of the chance constraint can be omitted.

 That is, given a 90% level of reliability on having sufficient storage at the end of any month t, two events are considered possible: the historical flow and the 10-percentile flow. The release chosen must accommodate the worst of these two flow situations to meet the level of reliability and achieve a nonnegative storage. If the 10-percentile flow is the smaller, release must be small enough to honor the probabilistic storage constraint as the binding constraint. If the historical flow is the smaller, release must be small enough that the next end-of-period storage is nonnegative. Then, no matter which of the constraints is binding, the next end-of-period storage is defined by the combination of reservoir capacity, the historical flow that actually occurs, and the chosen release.

Put another way, the historical record is thought of as forming the basis for planning the maximum steady release. At the end of each period of operation, however, the release must take into account not only the inflow that did occur but the inflow that could have occurred with the specified level of probability.

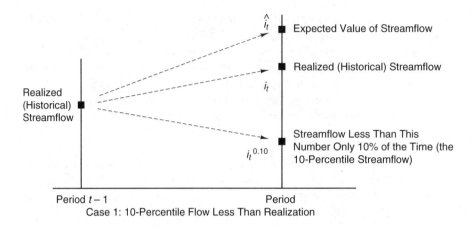

Case 1: 10-Percentile Flow Less Than Realization

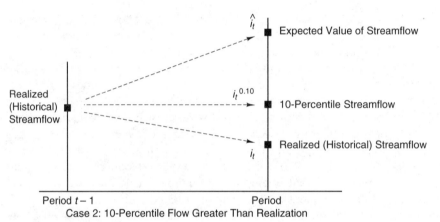

Case 2: 10-Percentile Flow Greater Than Realization

FIGURE 7-1   The binding constraint. In Case 1, if there is a binding constraint in period t, the deterministic equivalent of the chance constraint would be the binding constraint. In Case 2, the storage constraint that requires nonnegative storage would be the binding constraint.

Having criticized synthetic hydrology as a means for duplicating persistence, the question must be asked: "Does the proposed methodology do better?" We can argue that the proposed methodology does build in the effect of persistence in the calculation of the design or of maximum steady release. The effect is built in because the historical record and hence the critical period form not only the backbone of the procedure. In addition, during the sequence in which the persistently low flows occur, the methodology makes use of these low flows as the base for the calculation of the $(1 - \alpha)$-percentile flows. That is, the possibility of still lower flows occurring based on these lowest flows is considered as a distinct element of the procedure.

Depending on the flow values that are realized in the historical record, the flow in the upcoming month *could be* even lower than the value that actually

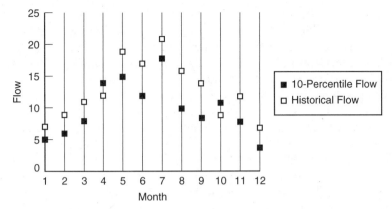

FIGURE 7-2 Historical flows and 10-percentile flows compared to a hypothetical case.

occurred. In particular, consider each month in which the flow in the persistently low grouping exceeds the conditional $(1 - \alpha)$-percentile flows (as opposed to the unconditional $(1 - \alpha)$-percentile flow). It is not clear in every situation that such months will occur. Occurrence depends on the parameters of serial correlation and on the chance constraint. The higher the required level of probability for nonnegative storage, the more likely such months (flow exceeds $[1 - \alpha]$-percentile flow) are to occur. In these months, the steady release is constrained by the conditional $(1 - \alpha)$-percentile flow, rather than the historically realized flow. The predicted $(1 - \alpha)$-percentile flows are based, however, on the flows from the cluster of persistently low flows. That is, we have potentially utilized the persistence in the record, piggy-backed off it, if you will, to create an even more constraining situation than occurred in the historical situation. It is in this sense that the procedure suggested here makes use of and builds in the persistence that may characterize droughts and critical periods.

The model presented here, however, is a design/operation model, not a predictive model. It does make use of a conditional predictive model that has previously been utilized in the generation of synthetic sequences. Thus, as structured here, the design/operation model may experience to some extent the same inadequacy that the conditional predictive models used in synthetic sequences experience, namely, a lack of ability to deliver flows of sufficiently low magnitude for a long enough duration. Because conditional low flows in the design model are based at every month on the historical record, however, that does display lengthy low-flow sequences, this possible flaw is mitigated to an extent. *It should be noted, moreover, that a better conditional model, when one is built, can be immediately incorporated into the design/operation model.* That is, the better estimates of the $(1 - \alpha)$-percentile flows would replace the estimates from the current conditional models.

In addition, in the examples to be presented here, the correlation structure used was exemplary only—in that correlations examine a future month's flow only as a function of the previous month's flow. To the extent that this conditional model is not sufficiently descriptive, the conditional flows will fall short of mimicking the correct conditional flow structure.

The methodology described here may be used to provide a trade-off curve between reliability and the maximum steady release, as we show subsequently. It can also be used to arrive at an estimate of the reliability associated with the maximum steady release that can be delivered from the historical record. That is, we can calculate the "safety" of the "safe" yield. The safe yield (historic yield is a better term) is calculated using the historical record as the base where a reservoir of given volumetric capacity is in place. In fact, the reliability of the safe yield is an end point of the trade-off curve we have just described. (It should be noted that the reliability we refer to here is that of providing storage in a nonnegative state, that is, not having to cut back on water delivery. In fact, when we refer to the probabilistic constraint on nonnegative storage, we mean that the probability that the release would not have to be reduced to prevent storage from hitting bottom is greater than α.)

The trade-off curve determination and the calculation of the reliability of safe yield proceed as follows. The $(1 - \alpha)$-percentile flow is associated with a constraint requiring the reliability of nonnegative end-of-period storage to be greater than or equal to α. The formula we used for the 10-percentile flow is trivially extended to the $(1 - \alpha)$-percentile flow as follows:

$$i^{1-\alpha} = (ax + b) - t_{1-\alpha}\, s_Y\, (1 - r^2)^{1/2}$$

where

x = flow recorded in the previous month
$t_{1-\alpha}$ = factor that translates the expected flow $(ax + b)$ to the $(1 - \alpha)$-percentile flow

The larger is α, the larger is $t_{1-\alpha}$.

Consider a situation in which the realized flow is less than the 10-percentile flow. We can plot $i^{1-\alpha}$ against α (see Figure 7-3) for this situation. Note that the realized flow corresponds to a flow at about the seven-percentile level, for this hypothetical example. For this situation, the historical flow provides the governing (tighter) constraint on storage.

The deterministic equivalent of the chance constraint on storage at the end of period t is, once again:

$$q - s_{t-1} \leq i_t^{1-\alpha}$$

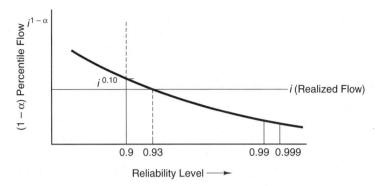

FIGURE 7-3 Reliability and flow level.

The constraint *implied* (but not written in this form) by the nonnegativity requirement on storage as calculated based on historical flows is

$$q + w_t - s_{t-1} \le i_t$$

As the reliability level α decreases from 0.999 to 0.9, the $(1 - \alpha)$-percentile flow increases. Given that the initial reliability level is 0.999, at first, the $(1 - \alpha)$-percentile flow binds, but, eventually, the historical flow takes over the binding role. Once the shift in binding constraint occurs for period t, further decreases in reliability level α have no effect. The same result obtains for each time period t—most likely at different levels of α. Each level of α produces its particular set of binding constraints and particular maximum steady release. Of course, for some ranges of α, the set of binding constraints does not change, but, nonetheless, the value of maximum steady release should change continuously (since the right-hand side changes continuously). When the set of binding constraints change, rates of change of the maximum steady release should change as well. The process produces a trade-off curve between maximum steady release and the reliability of reservoir storage (no shortage).

At the first value of α (as α is being decreased) at which the last of the $(1 - \alpha)$-percentile flows ceases to bind, the value of the maximum sustainable release stabilizes. At this point, all of the binding storage constraints are defined by the historical flows and not the $(1 - \alpha)$-percentile flows. Further changes in the reliability level α will cause no change in the binding constraints and, hence, no change in the maximum steady release. The trade-off curve is truncated. Nevertheless, at the point of truncation, important information is obtained. Consider the reliability level α at the point of truncation at which the last of the $(1 - \alpha)$-percentile flows bind. Note that the constraint will bind at this point in concert with an historical flow. At this point, this reliability level is the reliability of nonnegative storage associated with the maximum steady release derived from the historical flows alone. It is the reliability level associated with delivering the historic or safe yield.

THE DURATION OF A DROUGHT

The model used here for determining either required capacity or maximum sustainable yield opens further a question for consideration that has only recently begun to be examined, namely, the duration of a drought (Vogel and Bolognese, 1995). We showed earlier that the chance constraint on nonnegative storage

$$P[S_t \geq 0] \geq \alpha$$

can be converted to the deterministic equivalent

$$s_{t-1} \geq q - i_t^{1-\alpha}$$

An examination of this constraint reveals interesting issues.

Suppose α is thought of as approaching unity. That is, storage greater than zero is required with virtual certainty. As α approaches one, the $(1 - \alpha)$-percentile flow, $i_t^{1-\alpha}$ approaches zero. Thus, the right-hand side of the deterministic equivalent above approaches q. That is to say, a month-to-month maintenance of q, the monthly demand, as the base contents of the reservoir at the end of each month appears to ensure 100% reliability of surviving a drought. We know, however, that such a maintenance of storage will not provide 100% reliability. Maintaining q as end-of-month storage must be achieving something else, but what? And what does maintaining $(q - i_t^{1-\alpha})$ really achieve?

What maintaining $(q - i_t^{1-\alpha})$ as the base end-of-month storage is achieving is ensuring that the probability of surviving a drought *of one-month duration* is at least α. It is not the $(1 - \alpha)$-percentile drought of every duration that is protected for, only the one-month $(1 - \alpha)$-percentile drought and specifically the one-month drought that begins within the context of the historical record.

This leads the analyst to an additional constraint or an additional parameter or an additional measure of drought that must be specified for a meaningful statement of the problem of reservoir operation or design. Since there are already twin problems, smallest required capacity (Problem Ib) and maximum sustainable yield (Problem Ia), we state both problems in this new setting:

1. What is the smallest reservoir capacity c that will ensure a steady monthly release of q over a drought of k months duration with α reliability?

2. What is the largest steady release q that can be sustained given a reservoir capacity c over a drought of k months duration with α reliability?

We indicate the solution to this problem in the context of a drought of a specified length. We choose a one-year duration simply for illustration. Any length drought (2, 3, 4 month, etc.) and any reliability could be specified.

Suppose it is desired that the reservoir needs to be large enough to sustain release equal to demand over one year with α reliability given the context of the historical record. The $(1 - \alpha)$-percentile year-long drought could begin at any point in the recorded streamflow history and its depth will be directly related to the magnitude of the sequence of flows that directly preceded the onset of the drought.

We begin with the conceptual constraint:

P [sustaining release equal to demand throughout a year-long

drought that ends at time $t + 12$] $\geq \alpha$

The drought began 12 months earlier than $t + 12$, that is, at a time when the storage was at some realized level s_t. From the end of month t to the end of month $t + 12$, the total or sum of monthly inflows is represented by J_t, a random variable representing the cumulative flow in those 12 months. The probabilistic constraint becomes

$$P[s_t + J_t - 12_q \geq 0] \geq \alpha \qquad t = 1, 2, \ldots, n$$

In the preceding constraint, it is presumed that the releases are the same throughout the year. If they were not, but followed a cyclic pattern through the year, the constraint would read

$$P\left[s_t + J_t - \sum_{k=1}^{12} q_k \geq 0 \right] \geq \alpha$$

where q_k is the required release in month k of the year, $k = 1, 2, \ldots, 12$, a fixed proportion of the annual yield.

The values of the q_k may be specified or may be calculated based on an annual yield value Q_A. For now, we treat only the case of constant releases month to month. Following the sequence of mathematical operations described earlier, the deterministic equivalent of the chance constraint with month-to-month constant releases is

$$s_t \geq 12q - j_t^{1-\alpha}$$

where $j_t^{1-\alpha}$ is the value of the conditional cumulative streamflow in the 12 months following t that is exceeded with α reliability (i.e., the $[1 - \alpha]$-percentile flow).

To arrive at the conditional density needed to calculate the $(1 - \alpha)$-percentile annual flow requires a repetition of the steps needed to obtain the conditional density for the monthly flow. That is, for each month t, the regression equation is established for the cumulative flow in the next 12 months from $t + 1$ to $t + 12$ as a function of the cumulative flow in the preceding 12 months from $t - 11$ to t. This is accomplished by regressing the cumulative flow in the 12 months following t against the cumulative flow in the 12 months preceding and inclusive of t for all $t = 12, 13, \ldots, n$. There are, in fact, still just 12 regression equations—one for all January to December versus previous January to December, one for all February to January versus previous February to January, and so on. As before, since stream flow is log-normally distributed, the regression equation is estimated for the logs of the data. Of course, the correlation need not be a lag-one model; other lags are reasonable. The lag-one model is used for purposes of illustration only.

The full model of maximizing monthly yield given a capacity c, with a constraint on sustaining release throughout a year-long drought with α reliability, is

$$\text{Maximize} \quad z = q$$

$$\text{s.t.} \qquad s_t \leq c \qquad t = 1, 2, \ldots, n$$

$$s_t - s_{t-1} + q + w_t = i_t \qquad t = 1, 2, \ldots, n$$

$$12q - s_t \leq j_t^{1-\alpha} \qquad t = 12, 13, \ldots, n$$

$$s_n - s_o \geq 0$$

$$s_t \geq 0$$

$$q \geq 0$$

$$w_t \geq 0 \qquad t = 1, 2, \ldots, n$$

Of course, if q is given, the operative reliability constraint reads

$$s_t \geq 12q - j_t^{1-\alpha}$$

in which case capacity c is minimized.

For other durations of droughts where capacity is given and q is unknown, the constraints would look like the following:

For a two-month drought:

$$2q - s_t \leq (i2)_t^{1-\alpha} \qquad t = 2, 3, \ldots, n \tag{7-17}$$

For a three-month drought:

$$3q - s_t \leq (i3)_t^{1-\alpha} \qquad t = 3, 4, \ldots, n \tag{7-18}$$

For a four-month drought:

$$4q - s_t \leq (i4)_t^{1-\alpha} \qquad t = 4, 5, \ldots, n \qquad (7\text{-}19)$$

For a five-month drought:

$$5q - s_t \leq (i5)_t^{1-\alpha} \qquad t = 5, 6, \ldots, n \qquad (7\text{-}20)$$

and so on. In the preceding equations:

- $(i2)_t^{1-\alpha}$ is the conditional $(1 - \alpha)$-percentile flow cumulative over two months after month t (i.e., in months $t + 1$ and $t + 2$) *or* the conditional two-month flow that is exceeded with α reliability
- $(i3)_t^{1-\alpha}$ is the conditional $(1 - \alpha)$-percentile flow cumulative over three months after month t (i.e., in months $t + 1, t + 2, t + 3$) *or* the conditional three-month flow that is exceeded with α reliability; and so on

Computational experience with this model is provided later in the text. We will explore capacities needed to sustain yields through droughts of one, two, three months, and so on.

The recognition that the reliability constraint must contain a specification of a drought duration is both useful and potentially perplexing. On the one hand, it provides an explanation of the actual significance of the model's reliability constraints. This relatively satisfying explanation of the constraints, however, is supplanted by a new set of questions about reliability. Specifically, these questions are: (1) Which drought duration and which reliability are to be chosen for the model? (2) Should reliability be constrained for multiple-drought durations and multiple reliabilities and smallest capacity sought within this framework? (3) If multiple durations and reliabilities are used, what should the relation of required reliabilities and durations be?

To expand on these questions, the first question addresses the issue of which drought duration is most important and the reliability with which water is to be supplied over that duration. Is just one drought duration most important and how could a reliability be attached? Perhaps there is a single most important duration. Then a trade-off between yield and reliability might be generated for that drought duration, begging to some extent the question of the most appropriate reliability.

If more than one duration is important, should the probability of surviving each of the drought durations be the same? Or should the required probability of survival decline as the duration increases? In mathematical terms, the several drought duration constraints might read:

P [sustaining release through a one-month drought] $\geq \alpha_1$

P [sustaining release through a two-month drought] $\geq \alpha_2$

.

.

.

P [sustaining release through a k-month drought] $\geq \alpha_k$

What should the relation of $\alpha_1, \alpha_2, \ldots, \alpha_k$, be?

How can an analyst be expected to specify the appropriate reliability for each duration? Would one duration and reliability be expected to dominate other duration-reliability combinations? Answers were so much simpler with synthetic sequences. The computational experience on droughts of different durations suggests an ambiguous answer to the question of which duration is most constraining.

Two further comments can be offered about the utility of the model presented here. First, and critically important, the model is not tied to the predictive structure suggested by standard synthetic hydrology models. If another model of one or another econometric forms is created that mimics flows records better than these standard models, it should be able to be incorporated in this formulation and it should be used to derive $(1 - \alpha)$-percentile flows.

Second, whereas this model will find obvious use in assessing existing and planned water supply reservoirs, it should also find use in models of multipurpose reservoirs. Where reservoirs are designed or operated for multiple functions and a reliable water supply is one of those functions, there presently appears to be no generally applicable and efficient way to simultaneously optimize on all of the functions and enforce a reliability constraint on the water supply service. This model opens the possibility to optimize on multiple functions and enforce a stated level of reliability.

COMPUTATIONAL EXPERIENCE

Storage-Yield Curves

We have tested most of these models on historical streamflow data from the Gunpowder River in Maryland and from the North Branch of the Potomac, which rises in West Virginia. The results that follow indicate an expected order of behavior. The flow data was treated as log-normal, so that the conditional expected flow model utilized correlation of the logs of flow. The $(1 - \alpha)$-percentile flows were established by finding the $(1 - \alpha)$-percentile of log flows and taking the antilog.

The first step in the analysis was the calculation of classical storage-yield curves first for the Gunpowder River and then for the North Branch of the

Potomac. Two kinds of storage-yield curves may be calculated. The first type of curve is the garden variety storage-yield curve that is calculated using the historical record, and the historical record alone. The calculation was done using linear programming, but it could as well have been done using the methodology of sequent peak. The capacity-yield pairs based on the historic record are displayed in Figures 7-4 and 7- 5. Also indicated in the figures are the storage-yield curves for one-month-duration droughts where the reliability of nonnegative storage is at least 0.70 and 0.90.

The second kind of storage-yield curve consists of multiple curves and these are displayed in Figures 7-6 and 7-7 with data contained in Tables 7-1 and 7-2. These graphs consist of capacity-yield pairs in which end-of-month storages are constrained to be positive with 90% reliability, but the multiple curves

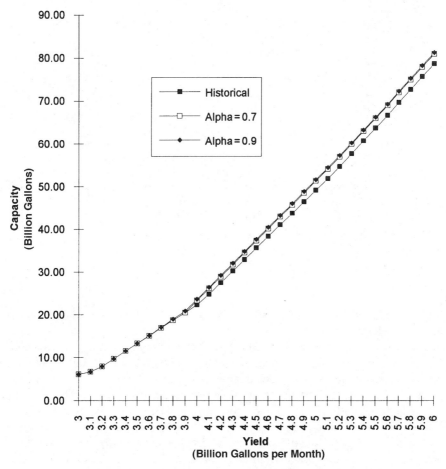

FIGURE 7-4   Capacity versus yield: Gunpowder River.

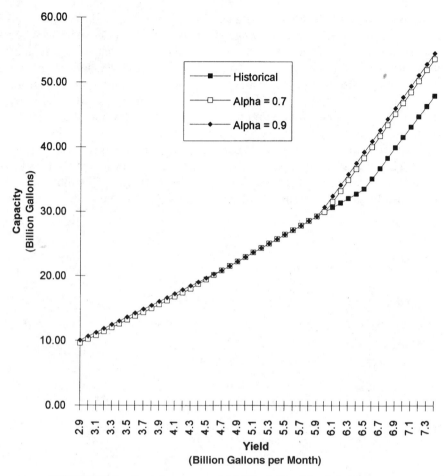

FIGURE 7-5 Capacity versus yield: North Branch of the Potomac.

reflect reliability constraints on having enough water in storage to meet monthly yields for one month, for two months, for three months, for four months, for 12 months, and for 24 months. Each curve has only one duration of drought associated with it, for example, the one-month drought, the two-month drought, and so on. The reliability level for all the curves was chosen arbitrarily at 0.90 to illustrate the type of results that could occur. Other levels of reliability could certainly have been chosen. The reliability levels could be the same for all durations of meeting demand as in this example, or they could be different for the different durations. The historic storage-yield curves are the same as the storage-yield curves for the 24-month duration drought.

The required capacity at each level of yield would be read as the upper or interior portion of the family of capacity-yield curves. That is, at every level

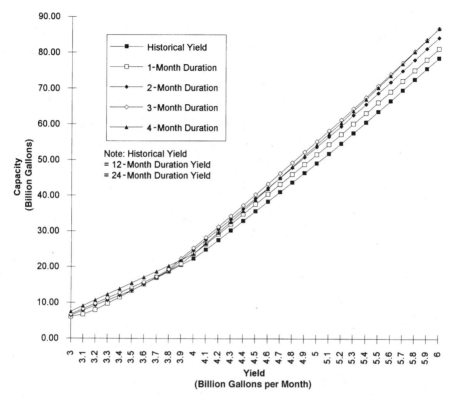

FIGURE 7-6 Capacity versus yield for varying durations: Gunpowder River (alpha = 0.90).

of yield, the required capacity is the highest of the capacities associated with all the durations under consideration. For Gunpowder River flows, which are analyzed in Figure 7-6, up to the level of 3.8 billion gallons per month, the four-month duration curve binds. At yield levels of 3.9 billion gallons per month and beyond, the three-month duration curve binds. We did not examine other durations because results are river-specific and reliability-specific and could also be different at other across-the-board reliability levels. In addition, because the conditional flow model was chosen for illustrative purposes only (i.e., was only a lag-one model), a different conditional flow model could provide different answers as well.

A parallel set of storage yield curves are calculated for the North Branch of the Potomac (see Figure 7-7 and Table 7-2). Once again, the capacity developed from historic flow data corresponded to the model that utilized the 24-month-duration constraints. And, as before, the upper or interior portion of the family of curves reflects required capacity as a function of yield level. In this case, at yield levels up to 5.6 billion gallons per month, the model with the two-month-duration constraints defines the required capacity. At 5.7 billion

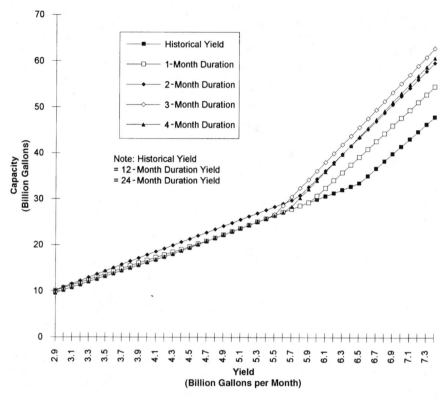

FIGURE 7-7 Capacity versus yield for varying durations: North Branch of the Potomac ($\alpha = 0.90$).

gallons per month and beyond, the model with the three-month constraints defines the required capacity levels. Of especial interest is the widening gap between the storage-yield curve based on the historic record (the 24-month-duration curve) and the defining storage-yield curve. For instance, at 6.4 billion gallons per month, the capacity required by the defining storage-yield curve is about one-third larger than the capacity based on the historic flow data alone.

Many more experimental runs could be done for different durations and reliabilities, but again, it must be stressed, the runs are meant to be exemplary rather than definitive. Of course, to determine the upper envelope, all drought durations should be included in one model so that capacity can be determined at a particular yield and reliability level without doing multiple runs. That is, the problem (7-1) to (7-7) should have appended to it constraints (7-17) to (7-20) as well as constraints for any other duration drought thought necessary to include. We did multiple runs here simply to show the relative effects of different durations.

In general, it will be necessary to do at least two math programming runs to establish precisely the interior portion of the family of curves, that is, the

TABLE 7-1 Storage-Yield Curve Data for the Gunpowder River

Yield	Historical Yield	1-Month Duration	2-Month Duration	3-Month Duration	4-Month Duration	12-Month Duration
3.0	6.12	6.12	6.45	6.76	7.51	6.12
3.1	6.72	6.72	7.85	8.26	9.11	6.72
3.2	7.93	7.96	9.25	9.76	10.71	7.93
3.3	9.73	9.73	10.65	11.26	12.31	9.73
3.4	11.53	11.53	12.05	12.76	13.91	11.53
3.5	13.33	13.33	13.45	14.26	15.51	13.33
3.6	15.13	15.16	15.13	15.76	17.11	15.13
3.7	16.93	17.06	16.93	17.26	18.71	16.93
3.8	18.73	18.96	18.94	19.31	20.31	18.73
3.9	20.53	20.86	21.84	22.31	21.91	20.53
4.0	22.33	23.62	24.74	25.31	23.69	22.33
4.1	24.83	26.42	27.64	28.31	26.69	24.83
4.2	27.53	29.22	30.54	31.31	29.69	27.53
4.3	30.23	32.02	33.44	34.31	32.69	30.23
4.4	32.93	34.82	36.34	37.31	35.78	32.93
4.5	35.63	37.62	39.24	40.31	38.88	35.63
4.6	38.33	40.42	42.14	43.31	41.98	38.33
4.7	41.03	43.22	45.04	46.31	45.08	41.03
4.8	43.73	46.02	47.94	49.31	48.18	43.73
4.9	46.43	48.82	50.84	52.31	51.28	46.43
5.0	49.13	51.62	53.74	55.31	54.38	49.13
5.1	51.86	54.42	56.64	58.31	57.48	51.86
5.2	54.66	57.22	59.54	61.31	60.58	54.66
5.3	57.65	60.21	62.63	64.50	63.87	57.65
5.4	60.65	63.21	65.73	67.70	67.17	60.65
5.5	63.65	66.21	68.83	70.90	70.47	64.71
5.6	66.65	69.21	71.93	74.10	73.77	68.81
5.7	69.65	72.21	75.03	77.30	77.07	72.91
5.8	72.65	75.21	78.13	80.50	80.37	77.01
5.9	75.65	78.21	81.23	83.70	83.67	81.11
6.0	78.65	81.21	84.33	86.90	86.97	85.21

$\alpha = 0.90$, varying durations.

correct storage-yield curve. As an example, the first run might maximize q subject to (7-1) to (7-7) plus (7-17) and (7-18). The run will produce a single storage-yield curve. The analyst will be unable to identify from this curve the binding drought durations, however. Then the problem would be solved again with constraint (7-19) added to the constraints of the first run.

If the storage-yield curve is exactly the same as without constraint (7-19), then the four-month drought introduces no new binding portion of the storage-yield curve. The problem is likely complete—although there is a remote possibility that a still longer duration drought could change the trade-off curve.

TABLE 7-2 Storage-Yield Curve Data for the Potomac River, North Branch

Yield	Historical Yield	1-Month Duration	2-Month Duration	3-Month Duration	4-Month Duration	12-Month Duration
2.9	9.60	10.06	10.13	9.60	9.60	9.60
3.0	10.20	10.66	10.83	10.20	10.20	10.20
3.1	10.80	11.26	11.53	10.80	10.80	10.80
3.2	11.40	11.86	12.23	11.40	11.40	11.40
3.3	12.00	12.46	12.93	12.00	12.00	12.00
3.4	12.60	13.06	13.63	12.60	12.60	12.60
3.5	13.20	13.66	14.33	13.20	13.20	13.20
3.6	13.80	14.26	15.03	13.80	13.80	13.80
3.7	14.40	14.86	15.73	14.40	14.40	14.40
3.8	15.00	15.46	16.43	15.00	15.00	15.00
3.9	15.60	16.06	17.13	15.60	15.60	15.60
4.0	16.20	16.66	17.83	16.20	16.20	16.20
4.1	16.80	17.26	18.53	16.80	16.80	16.80
4.2	17.40	17.86	19.23	17.40	17.40	17.40
4.3	18.03	18.46	19.93	18.03	18.03	18.03
4.4	18.73	19.06	20.63	18.73	18.73	18.73
4.5	19.43	19.66	21.33	19.43	19.43	19.43
4.6	20.13	20.26	22.03	20.13	20.13	20.13
4.7	20.83	20.86	22.73	20.83	20.83	20.83
4.8	21.53	21.53	23.43	21.53	21.53	21.53
4.9	22.23	22.23	24.13	22.23	22.23	22.23
5.0	22.93	22.93	24.83	22.93	22.93	22.93
5.1	23.63	23.63	25.53	23.63	23.63	23.63
5.2	24.33	24.33	26.23	24.33	24.33	24.33
5.3	25.03	25.03	26.93	25.03	25.03	25.03
5.4	25.73	25.73	27.63	25.73	25.73	25.73
5.5	26.43	26.43	28.33	26.68	26.43	26.43
5.6	27.13	27.13	29.03	28.58	27.13	27.13
5.7	27.83	27.83	29.73	30.48	28.41	27.83
5.8	28.53	28.53	30.84	32.38	30.31	28.53
5.9	29.23	29.23	32.64	34.28	32.21	29.23
6.0	29.93	30.70	34.44	36.18	34.11	29.93
6.1	30.63	32.40	36.24	38.08	36.01	30.63
6.2	31.33	34.10	38.04	39.98	37.91	31.33
6.3	32.03	35.80	39.84	41.88	39.81	32.03
6.4	32.73	37.50	41.64	43.78	41.71	32.73
6.5	33.50	39.20	43.44	45.68	43.61	33.50
6.6	35.10	40.90	45.24	47.58	45.51	35.10
6.7	36.70	42.60	47.04	49.48	47.41	36.70
6.8	38.30	44.30	48.84	51.38	49.31	38.30
6.9	39.90	46.00	50.64	53.28	51.21	39.90
7.0	41.50	47.70	52.44	55.18	53.11	41.50
7.1	43.10	49.40	54.24	57.08	55.01	43.10
7.2	44.70	51.10	56.04	58.98	56.91	44.70
7.3	46.30	52.80	57.84	60.88	58.81	46.30
7.4	47.90	54.50	59.64	62.78	60.71	47.90

$\alpha = 0.90$, varying durations.

If the storage-yield curve does change when constraint (7-19) is added, then the analyst should run the problem once more with constraint (7-20) appended to see if the five-month drought changes the picture.

The sequence of runs described, although it can decrease total optimization computation and analyst effort to establish the appropriate storage-yield curve, does not identify those portions of the storage-yield curve in which the two-month drought applies, the three-month drought applies, the four-month drought applies, and so on. If this information is deemed important, then the analyst should carry out the incremental analysis of producing the storage-yield curves for the two-month drought, the three-month drought, and so on, and overlaying all of the curves on a single graph as we did in Figures 7-6 and 7-7 for the Gunpowder River and the North Branch of the Potomac.

Yield-Reliability Curves

We indicated earlier that by an investigation of the trade-off between yield and reliability for a given capacity reservoir, we could begin the process of determining the actual reliability of the historic yield. By historic yield, we mean the maximum steady release achievable through the historic flow record. First, we perform the calculations here for the model that uses a one-month duration drought, again as an example of the kinds of calculation that can be made. That is, since other duration constraints could be utilized, we cannot regard the result as definitive. Nonetheless, the results provide much interesting information.

For the Gunpowder River, we examine curves of reliability versus yield for reservoir capacities of 15 and 20 billion gallons (Figures 7-8 and 7-9). The model maximizes yield given the capacity and a stated reliability level and solves repeatedly as the reliability level is decreased. Obviously, the yield associated with a capacity level will increase as the required level of reliability is reduced. As pointed out earlier, at some level of reliability, yield no longer increases. This is represented by the flat spot on Figure 7-8. At the level at which yield no longer increases, the historic flow record binds. Hence, one may interpret the reliability associated with this yield level as the reliability of the historic yield—*if* the one-month duration model is the best model for determining capacity. In Figure 7-8, for a reservoir capacity of 15 billion gallons, it appears that the reliability associated with the safe yield is about 0.89. Figure 7-9 displays the yield-reliability curve for a capacity of 20 billion gallons. The reliability of the historic yield associated with this capacity is read as 0.73.

Two curves of yield versus reliability are also shown for the North Branch of the Potomac—at reservoir capacities of 20 and 30 billion gallons (Figures 7-10 and 7-11). These curves reach their respective flat spots, that is, the reliability levels at which the historic flow data binds at 85% (yield = 5.808 billion gallons/month), and at 70% (yield = 6.0094).

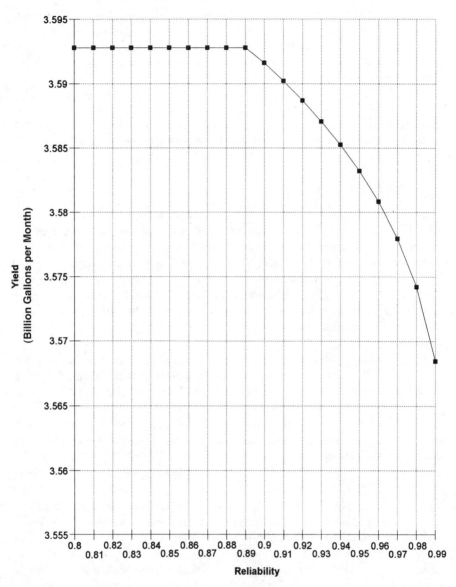

FIGURE 7-8 Yield versus reliability: Gunpowder River (capacity = 15 billion gallons).

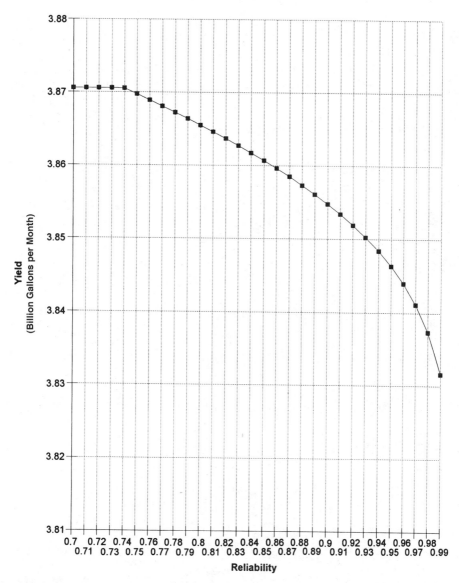

FIGURE 7-9   Yield versus reliability: Gunpowder River (capacity = 20 billion gallons).

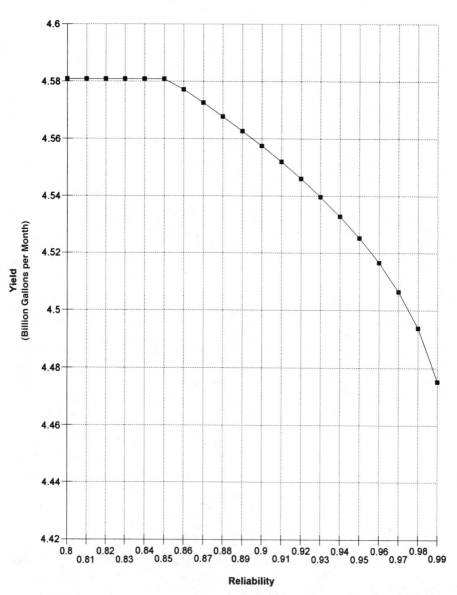

FIGURE 7-10   Yield versus reliability: North Branch of the Potomac (capacity = 20 billion gallons).

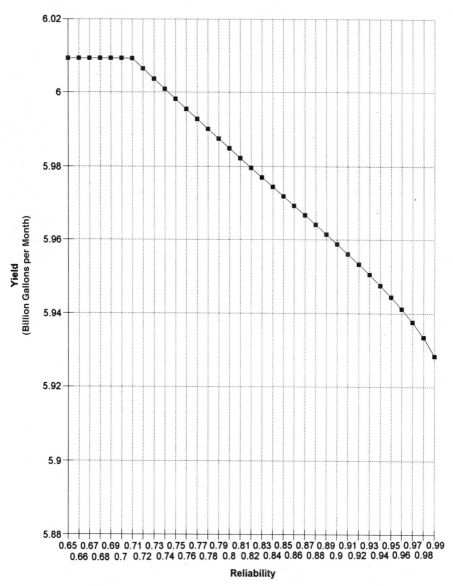

FIGURE 7-11   Yield versus reliability: North Branch of the Potomac (capacity = 30 billion gallons).

We also calculated capacity versus annual yield curves for the Gunpowder and the North Branch (see Figures 7-12 and 7-13). For these curves, the values of β, the proportion of annual yield required in a given month, in order from January to December were 0.05, 0.05, 0.05, 0.05, 0.09, 0.12, 0.14, 0.14, 0.12, 0.09, 0.05, and 0.05. The curves for the historic storage-yield relationship as well as for reliabilities of 0.70 and 0.90 are all shown.

The final example to be presented is yet another analysis of the reliability of the historic yield. Earlier, we examined the trade-off between reliability and yield for the Gunpowder and North Branch of the Potomac where the drought duration was taken as a single month (and a lag-one relation was used to create conditional densities). As required reliability was decreased, yield could increase until the mass balance constraints associated with the historic flows became binding. Below this level of reliability, yield did not change. Hence, the implied reliability of the historic yield was the reliability at which historic flow records became binding

We now repeat the analysis for the North Branch of the Potomac using a capacity of 20 billion gallons and a drought duration of two months. The correlation relationship used to create the appropriate conditional density is now two months versus two months. When we back off on the reliability constraint (i.e., reduce the required reliability), we find that the historic flow

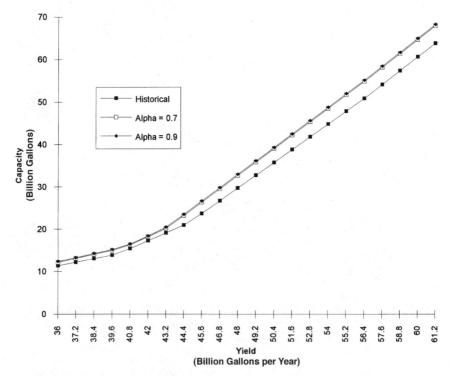

FIGURE 7-12 Capacity versus annual yield: Gunpowder River.

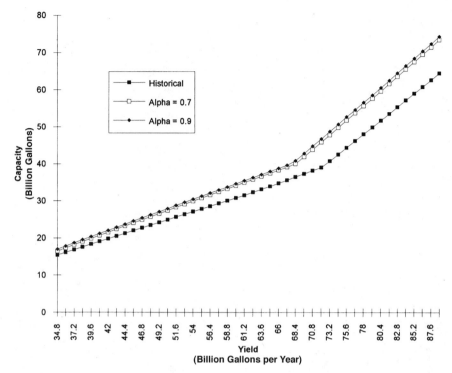

FIGURE 7-13   Capacity versus annual yield: North Branch of the Potomac.

record binds at a reliability of 0.62 (Figure 7-14) rather than the reliability level of 0.85 obtained when we considered a one-month drought. The result illustrates that examination with other drought durations may well be necessary to establish an estimate of the actual reliability of the historic yield.

SUMMARY

The determination of the "safe yield" of a reservoir used only for water supply is a classic problem whose solution has been approached incrementally through time as mathematical techniques have evolved. In the 1960s, mass curve analysis was replaced by "sequent peak" and by linear programming, but both techniques were designed to operate on historic records, making the reliability of the calculated yield problematic. With the introduction of the methodology of synthetic hydrology, synthetic sequences could be created whose statistics were almost comparable to those of the historic record. Testing a yield value on a number of synthetic records made possible the calculation of a rough measure of the reliability achieved by a reservoir of given capacity.

This chapter combines the ideas of synthetic hydrology with the power of chance constraints to specify required reliabilities of service. Using the syn-

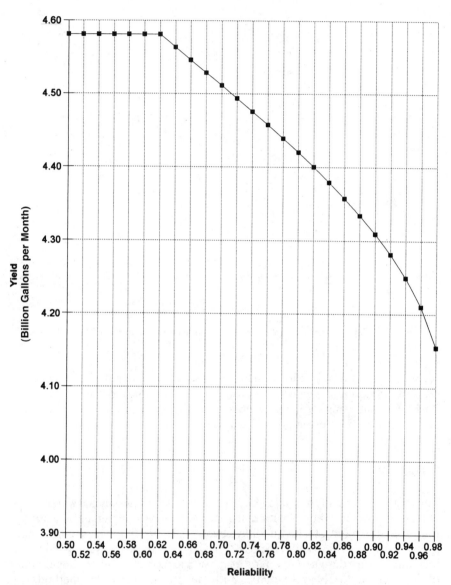

FIGURE 7-14   Yield versus reliability for two-month duration flow: North Branch of the Potomac (capacity = 20 billion gallons).

thetic hydrology model to derive the conditional densities of stream flows allows derivation of $(1 - \alpha)$-reliable flow events, within the context of flows that occurred in the historical record. Thus, explicit constraints can be placed on the reliability of nonnegative storage, allowing circumvention of the use of synthetic sequences. The reservoir model utilizes chance constraints that are conditioned on previous flow(s) using the aforementioned densities. Example calculations of storage-yield curves are offered for droughts of various durations. The model is extended conceptually to multiple water supply reservoirs on parallel streams, but it is also easy to envisage extension to the single multiple-purpose reservoir as well as to a number of multiple-purpose reservoirs in parallel or in other arrangements. Nonetheless, the complete extension to multiple reservoirs presents challenging mathematical issues.

ACKNOWLEDGMENT

I would like to acknowledge assistance in programming of the example problems from Stephanie Snyder and Joseph Harkness. They were wonderfully helpful and patient in demanding tasks.

APPENDIX: DERIVATION OF THE VARIANCE OF THE RANDOM SHOCK

The random-shock model suggests, in the lag-one model, that this month's flow, Y, is a linear function of last month's flow, X, modified by a flow shock, R, that moves this month's flow away from its expected value:

$$Y = aX + b + R \qquad (7A\text{-}1)$$

That is, $aX + b$ is the expected flow this month, and R, a number that may be positive or negative, shifts the occurring value away from its expectation. The random shock has an expected value, μ_R, of 0 and variance, σ_R^2, that require estimation.

Taking the expected value of Y yields

$$\mu_Y = a\mu_x + b \qquad (7A\text{-}2)$$

subtracting this equation from (7A-1) provides

$$Y - \mu_Y = a(X - \mu_X) + R \qquad (7A\text{-}3)$$

The covariance of Y with X is

$$\text{Cov}(Y, X) = \sigma_{Y,X} = E[(Y - \mu_Y)(X - \mu_X)]$$

or

$$\sigma_{Y,X} = E[(a(X - \mu_X) + R)(X - \mu_X)]$$
$$= aE[(X - \mu_X)^2]$$
$$= a\sigma_X^2 \qquad (7A\text{-}4)$$

Since the correlation coefficient is

$$\rho_{Y,X} = \frac{\sigma_{Y,X}}{\sigma_Y \sigma_X} \qquad (7A\text{-}5)$$

the value of a is given by

$$a = \rho_{Y,X} \frac{\sigma_Y}{\sigma_X} \qquad (7A\text{-}6)$$

and b, from Equation (7A-2), is

$$b = \mu_Y - \rho_{Y,X} \frac{\sigma_Y}{\sigma_X} \mu_X \qquad (7A\text{-}7)$$

so that (7A-1) becomes

$$Y = \rho_{Y,X} \frac{\sigma_Y}{\sigma_X} X + \left(\mu_Y - \rho_{Y,X} \frac{\sigma_Y}{\sigma_X} \mu_X\right) + R \qquad (7A\text{-}8)$$

The variance of Y, using Equation (7A-3), is

$$\sigma_Y^2 = E[(Y - \mu_Y)^2] = E[(a(X - \mu_X) + R)^2]$$
$$= a^2 \sigma_X^2 + \sigma_R^2 \qquad (7A\text{-}9)$$

Solving for the variance of the random shock provides

$$\sigma_R^2 = \sigma_Y^2 - \left(\rho_{Y,X} \frac{\sigma_Y}{\sigma_X}\right)^2 \sigma_x^2$$
$$= \sigma_Y^2 - \rho_{Y,X}^2 \sigma_Y^2$$
$$= \sigma_Y^2 (1 - \rho_{Y,X}^2)$$

Hence, the sample estimate of the variance of the random shock is

$$s_R^2 = s_Y^2 (1 - r^2)$$

where s_Y^2 is the sample estimate of the variance of Y, and r^2 is the sample estimate of the correlation coefficient.

REFERENCES

Charnes, A., and W. Cooper. 1959. "Chance Constraint Programming." *Management Science,* Vol. 6, pp. 73–79.

Fair, G., and J. Geyer. 1967. *Elements of Water Supply and Wastewater Disposal.* John Wiley, New York.

Fiering, M. 1964. "Multi-variate Techniques for Synthetic Hydrology." *Journal of the Hydraulics Division (ASCE),* Vol. 90, pp. 43–60.

Fiering, M. 1967. *Synthetic Hydrology,* Harvard University Press, Cambridge.

Fiering, M., and B. Jackson. 1971. *Synthetic Stream Flows,* American Geophysical Union, Washington, D.C. Water Resources Monograph 1.

Hirsch, R. 1978. *Risk Analyses for a Water Supply System—Occoquan Reservoir, Fairfax and Prince William Counties, Virginia,* Open File Report 78-452. U.S. Geological Survey, Reston, Va.

Houck, M., Cohon, J., and C. ReVelle. 1980. "Linear Decision Rule in Reservoir Management and Design, 6: Incorporation of Economic Efficiency Benefits and Hydroelectric Energy Generation," *Water Resources Research,* Vol. 16, pp. 196–200.

Moy, W.-C., J. Cohon, and C. ReVelle. 1986. "A Programming Model for Analysis of the Reliability, Resilience, and Vulnerability of a Water Supply Reservoir." *Water Resources Research,* Vol. 22, pp. 489–498.

ReVelle, C., and J. Gundelach. 1975. "Linear Decision Rule in Reservoir Management and Design, 4, A Rule that Minimizes Output Variance." *Water Resources Research,* Vol. 11, pp. 197–203.

ReVelle, C., E. Joeres, and W. Kirby. 1969. "Linear Decision Rule in Reservoir Management and Design, 1, Development of the Stochastic Model." *Water Resources Research,* Vol. 5, pp. 767–777.

ReVelle, C., and W. Kirby. 1970. "Linear Decision Rule in Reservoir Management and Design, 2, Performance Optimization." *Water Resources Research,* Vol. 6, pp. 1033–1044.

Rippl, W. "The Capacity of Storage Reservoirs for Water Supply." *Proceedings Institution of Civil Engineers,* Vol. 71, pp. 270–278.

Thomas, H., and R. Burden. 1963. *Operations Research in Water Quality Management,* Report to the Public Health Service. Harvard Water Resources Group, Harvard University, Cambridge.

Vogel, R., and R. Bolognese. 1995. "Storage-Reliability-Resilience-Yield Relations for Over-Year Water Supply Systems." *Water Resources Research,* Vol. 31, No. 3, pp. 645–654.

CHAPTER 8

REVISITING ALLOCATION WITH THE NEW MODEL ENFORCING A RELIABILITY REQUIREMENT

INTRODUCTION

In the first main section of Chapter 6, four models were structured to allocate reservoir services between water supply and flood control. Two of these allowed a flood pool that was variable by month. In these two models, we saw that requiring a stated level of reliability for the water supply function was a task with an extremely large computational burden—if it could be accomplished at all. The use of multiple synthetic sequences to establish the level of water supply reliability for a particular beginning set of 12 flood pool levels would have been followed by a 12-dimensional search to establish locally optimal values of the flood pool levels. Clearly, the task of using synthetic hydrology to find optimal flood pool volumes given a reliability requirement is enormous. This level of difficulty motivated the research reported in Chapter 7. In the second main section of Chapter 6, we also showed how allocation could be accomplished among water supply, flood control, and hydroelectric energy production.

In Chapter 7, a model for the operation or design of a water-supply-only reservoir was created that explicitly constrained to a tolerable level the probability of the reservoir's failure. The model accounted for the serial correlation of reservoir inflows by establishing conditional density functions for flow in the month(s) following recorded historical flow(s). These conditional densities were then used in chance constraints on nonnegative reservoir storage. These constraints were written within the context of operation through the historical record. It was recognized that nonnegative storage is the requirement that ensures continued operation at the designated water supply delivery levels.

In this chapter, the allocation of reservoir services is examined in two phases in the context of the new model for water supply reservoir operation or design. That new model allows the analyst to place explicit constraints on the reliability with which a water supply can be maintained. The two phases are the consideration first of water supply and flood control (in the first main section) and then of water supply, flood control, and hydropower (in the second main section).

In the first main section (water supply and flood control), three cases are examined: (1) constant monthly draft and constant flood pool through the year, (2) variable monthly draft and constant flood pool through the year, and (3) constant monthly draft and variable flood pool by month. The fourth case, variable monthly draft and variable flood pool, becomes easy to structure and so is not presented. An economic objective of maximizing value consists of three terms: revenue from the sale of water less flood losses less the annualized cost of the reservoir. The model presented here uses the new model for the water supply reservoir within an economic model that optimally allocates storages between water supply and flood control.

In the second main section, the new model, which constrains water supply reliability, is used to formulate the allocation of reservoir services among water supply, flood control, and hydropower. Only a single situation consisting of constant water supply demand, constant flood pool, and steady hydropower production is examined, but the extension to variable demands and variable flood pools is not difficult.

RELIABILITY CONSTRAINED MODELS FOR ALLOCATING WATER SUPPLY AND FLOOD STORAGE

Model 8-a: Constant Release and Constant Flood Pool

As before, we wish to allocate reservoir services to maximize an objective consisting of the annual revenue from the sale of water less flood losses less the amortized annual cost of the reservoir. On this occasion, we append reliability constraints. The earlier model was, in non-standard form:

$$\text{Maximize} \quad z = 12p_w q - f(v) - g(c) \tag{8-1}$$

$$\text{s.t.} \quad s_n \geq s_o \tag{8-2}$$

$$s_t = s_{t-1} - q - w_t + i_t \quad t = 1, 2, \ldots, n \tag{8-3}$$

$$s_t \leq c - v \quad t = 1, 2, \ldots, n \tag{8-4}$$

$$s_t, w_t \geq 0 \quad t = 1, 2, \ldots, n$$

$$q, c, v \geq 0$$

where p_w must be specified for solution of the problem.

The reliability constraints for a one-month duration drought are

$$P[S_t \geq 0] \geq \alpha \qquad t = 1, 2, \ldots, n$$

By using the mathematics assembled earlier, this constraint on surviving the one-month drought has a deterministic equivalent of

$$q - s_{t-1} \leq i_t^{1-\alpha} \qquad t = 2, 3, \ldots, n \qquad (8\text{-}5)$$

where $i_t^{1-\alpha}$ is the $(1 - \alpha)$-percentile conditional one-month flow in month t. We let

$(i2)_t^{1-\alpha}$ = conditional $(1 - \alpha)$-percentile two-month flow after month t; that is, a flow that begins with month $(t + 1)$'s flow

$(i3)_t^{1-\alpha}$ = conditional $(1 - \alpha)$-percentile three-month flow after month t; that is, a flow that begins with month $(t + 1)$'s flow

$(i4)_t^{1-\alpha}$ = conditional $(1 - \alpha)$-percentile four-month flow after month t; that is, a flow that begins with month $(t + 1)$'s flow

Then we can add to the one-month constraints (8-5), the following constraints:

$$2q - s_t \leq (i2)_t^{1-\alpha} \qquad t = 2, 3, \ldots, n \qquad (8\text{-}6)$$

$$3q - s_t \leq (i3)_t^{1-\alpha} \qquad t = 2, 3, \ldots, n \qquad (8\text{-}7)$$

$$4q - s_t \leq (i4)_t^{1-\alpha} \qquad t = 2, 3, \ldots, n \qquad (8\text{-}8)$$

From the investigations of Chapter 7, we know that we are unable to predict in advance which duration drought will bind since we do not know in advance the value of the water supply capacity under which we will operate. Thus, we should include the chance constraints for all possible reasonable drought durations that we could consider as possibly binding. Of course, some or one drought-duration constraint will bind and whichever one(s) does (do), it will have been included and we can identify it.

Model 8-b: Differing Monthly Releases and Constant Flood Pool

The objective of this model involves all of the same elements as the previous model except that now revenue is calculated as price times the annual amount of water sold. The model, in nonstandard form with its deterministic equivalents of the chance constraints on storage, is

Maximize $z = p_w Q_A - f(v) - g(c)$

s.t. $s_n \geq s_o$

$$s_t = s_{t-1} - q_k - w_t + i_t \quad \begin{array}{l} t = 1, 2, \ldots, n; \\ k = t - 12[(t-1)/12] \end{array} \quad (8\text{-}9)$$

$$s_t \leq c - v \quad t = 1, 2, \ldots, n \quad (8\text{-}10)$$

$$q_k - \beta_k Q_A = 0 \quad k = 1, 2, \ldots, 12 \quad (8\text{-}11)$$

$$s_t, w_t \geq 0 \quad t = 1, 2, \ldots, n$$

$$q_k \geq 0 \quad k = 1, 2, \ldots, 12$$

$$Q_A, c, v \geq 0$$

plus

$$q_k - s_{t-1} \leq i_t^{1-\alpha} \quad \begin{array}{l} t = 2, 3, \ldots, n; \\ k = t - 12[(t-1)/12] \end{array} \quad (8\text{-}12)$$

$$q_{k+1} + q_{k+2} - s_t \leq (i2)_t^{1-\alpha} \quad \begin{array}{l} t = 2, 3, \ldots, n; \\ k = t - 12[(t-1)/12] \end{array} \quad (8\text{-}13)$$

$$q_{k+1} + q_{k+2} + q_{k+3} - s_t \leq (i3)_t^{1-\alpha} \quad \begin{array}{l} t = 3, 4, \ldots, n; \\ k = t - 12[(t-1)/12] \end{array} \quad (8\text{-}14)$$

$$\sum_{r=k+1}^{k+4} q_r - s_t \leq (i4)_t^{1-\alpha} \quad \begin{array}{l} t = 4, 5, \ldots, n; \\ k = t - 12[(t-1)/12] \end{array} \quad (8\text{-}15)$$

As in the previous model, chance constraints associated with all drought durations that could cause binding are included in the model since it is not possible to predict the value of the water supply volume that will be chosen by the optimization.

Next we structure the first of the variable flood pool models.

Model 8-c: Constant Monthly Release and Variable Flood Pool

Once again, the model seeks to maximize the difference between revenues from the sale of water and the sum of annual flood losses and the amortized reservoir costs subject to the constraints in nonstandard form.

Maximize $z = 12 p_w q - \left(\sum_{k=1}^{12} f_k(v_k) + g(c) \right)$

s.t. $s_n \geq s_o$ (8-16)

$$s_t = s_{t-1} + i_t - q - w_t \quad t = 1, 2, \ldots, n \quad (8\text{-}17)$$

$$s_t \leq c - v_k \qquad t = 1, 2, \ldots, n;$$
$$k = t - 12[(t - 1)/12] \qquad (8\text{-}18)$$

$$s_t, w_t \geq 0 \qquad t = 1, 2, \ldots, n$$

$$v_k \geq 0 \qquad k = 1, 2, \ldots, 12$$

$$q, c \geq 0$$

Again, we add the deterministic equivalents of the chance constraints on nonnegative reservoir storages:

$$q - s_{t-1} \leq i_t^{1-\alpha} \qquad t = 2, 3, \ldots, n \qquad (8\text{-}19)$$

$$2q - s_t \leq (i2)_t^{1-\alpha} \qquad t = 2, 3, \ldots, n \qquad (8\text{-}20)$$

$$3q - s_t \leq (i3)_t^{1-\alpha} \qquad t = 3, 4, \ldots, n \qquad (8\text{-}21)$$

$$4q - s_t \leq (i4)_t^{1-\alpha} \qquad t = 4, 5, \ldots, n \qquad (8\text{-}22)$$

In this model, we will not, as was the case in Chapter 2, be operating at any particular water supply reservoir capacity but, instead, at a range of capacities. The water supply reservoir in month k will have a capacity of $(c - v_k)$. Now it might be thought that if the process were a reallocation of the services of an existing reservoir, that the water supply capacity would be known. In thinking this, one would suppose that only one such constraint need be added for each month, namely, the duration constraint that is binding. Of course, this is not the case, first, because the height of the water supply pool is still not known and, second, because the water supply pool will be variable by month. Writing the constraints for all reasonable drought durations that might bind may be a bit more work, but the answer without that effort may not be correct.

Discussion

The fourth model, differing monthly releases and variable flood pool, could be written, but does not offer much new to say.

Nevertheless, there is more to say about these allocation models that divide services between water supply and flood control with formal constraints on the reliability of water supply services. It is of interest that the reliability constraints are essentially outside the economic optimization portion of the model. Certainly, they do influence the answer, but no economic value is placed explicitly on reliability.

Wurbs and Cabezas (1987) did conduct a reallocation study of a reservoir in Texas in which economic losses associated with water shortages were included in the evaluation. However, we have not yet built a mechanism to take an accounting of shortages. All this model does is constrain the reliability of being able to meet a planned water supply release. We have not measured actual shortage quantities, although this may well be possible using the event responsive density adjustment methodology.

We can, however, approach the economic value associated with the reliability constraint by classical sensitivity analysis. Since each level of reliability has associated with it an allocation between water supply and flood control with an economic value as well, we can evaluate the trade-off between the value of the objective and the level of reliability. Obviously, the greater the reliability required, the more storage volume and economic value are shifted to water supply. The shape of the curve of objective value versus reliability promises to be informative.

As we discussed in Chapter 6, price p_w is varied to develop a supply curve of price versus monthly quantity. The supply curve shows the yield q that would be provided as a function of the price of water p_w. Nonetheless, each point on the curve also corresponds to a particular size reservoir and a division of reservoir services between water supply and flood control. That is, for each p_w, we would read a value of q from the graph, but there would be, in addition, a (set of) flood pool(s) and a reservoir capacity, c.

WATER SUPPLY, FLOOD CONTROL, AND HYDROPOWER

In this section, we add the hydropower function to the allocation process so that we are now sizing a reservoir, determining water supply release(s), the flood pool(s), and a hydroenergy firm supply level.

As in Chapter 6, we assume that the water released through the turbines for hydrogeneration becomes unavailable for water supply. We also assume that only firm energy is valued, which is the assumption with which we began the second main section of Chapter 6. In Chapter 6, we did relax that assumption, but accomplishment of the relaxation should be evident, so it will not be shown here.

The variables and parameters needed to treat the hydropower function are:

x_t = release through the turbines in month t

h_t = height of the water surface above the turbines at the end of month t

d = value of minimal monthly hydroenergy production, over which we will iterate

h_o, m = known intercept and slope of the function approximating the head versus storage curve

p_F = price of firm hydroelectric energy, known

p_w = unknown price users may be willing to pay for water

α = energy constant that converts to hydroenergy units after efficiency losses

As in Chapter 6, the plan is to iterate over values of firm power, d, to find the optimal division of reservoir services. The problem statement, without the value of hydropower included in the objective is, in nonstandard form:

$$\text{Maximize} \quad z = 12p_w q - f(v) - g(c) \tag{8-23}$$

$$s_n \geq s_o \tag{8-24}$$

$$s_t = s_{t-1} - q - x_t - w_t + i_t \qquad t = 1, 2, \ldots, n \tag{8-25}$$

$$\alpha \left[h_o + (m/2)s_t + (m/2)s_{t-1} \right] x_t \geq d \qquad t = 1, 2, \ldots, n \tag{8-26}$$

$$s_t \leq c - v \tag{8-27}$$

$$q - s_{t-1} \leq i_t^{1-\alpha} \qquad t = 2, 3, \ldots, n \tag{8-28}$$

$$2q - s_t \leq (i2)_t^{1-\alpha} \qquad t = 2, 3, \ldots, n \tag{8-29}$$

$$3q - s_t \leq (i3)_t^{1-\alpha} \qquad t = 3, 4, \ldots, n \tag{8-30}$$

$$4q - s_t \leq (i4)_t^{1-\alpha} \qquad t = 4, 5, \ldots, n \tag{8-31}$$

The constraints are interpreted as in Chapter 6 except for (8-28) to (8-31). Constraint (8-26) is the non-linear hydropower constraint that needs to be linearized. Division of both sides by x_t provides a form d/x_t that is convex and can be piecewise approximated. Because of the sense of the constraints, the piecewise variables enter in the correct order. Details are provided in Chapter 3 on hydropower. Constraints (8-28) to (8-31) are the set of deterministic equivalents of the drought-duration constraints that might be binding in the analysis of the reservoir's potential services. Parameters are as defined earlier in the chapter.

For any particular values of d, and price of water p_w, the problem would be solved. The price of water, p_w, is held constant while d is varied from small to large. For each value of d and each corresponding objective value (8-23), the sum of $12p_F d$, the annual revenue from the sale of firm hydroelectric energy, and the objective (8-23) is tabulated into a combined objective. The process is repeated over many values of d. The maximum of the many combined objective values is taken as the best solution with the optimal division of reservoir services between the three functions—for the specified value of p_w. The value of q associated with this best combined objective value over all d for a given p_w is placed as a point on the curve of q versus p_w.

Now, for a new value of p_w, the process of specifying multiple values of d is repeated. Once again, the objective value is calculated for each d that is given and a new objective tabulated that is the previously calculated term plus the annual value of hydroenergy, namely, $12p_F d$. This new objective value is placed on the curve of net value versus d. The maximum combined objective is located over all d values given, and the value of q associated with that solution is placed on the curve of q versus p_w. In such a way, the supply curve for water (q versus p_w) is built up.

This supply curve is slightly different in character than the supply curve discussed in the first main section of this chapter. That supply curve of q versus p_w revealed for each price p_w, not only the yield q, but also a reservoir capacity c and a flood pool v or set of flood pools v_k. This supply curve furnishes one more piece of critical information, the firm yield of hydroelectric energy d. Of course, d is not shown on the graph of q versus p_w, but each point on that curve has associated with it not only q and p_w, but also a capacity c, a flood pool v, and firm level of hydroenergy d.

We do not carry the analysis to dump energy valuation nor to the determination of storage-rule curves, but the information to do so is in Chapter 6.

REFERENCE

Wurbs, R., and L. Cabezas. 1987. "Analysis of Reservoir Storage Reallocations," *Journal of Hydrology,* Vol. 92, pp. 77–95.

CHAPTER 9

THE REALLOCATION OF RESERVOIR SERVICES USING THE RELIABILITY-CONSTRAINED RESERVOIR MODEL

INTRODUCTION

In Chapter 8, we showed how the new model for reservoir operation and design that was introduced in Chapter 7 could be utilized in the context of reservoir allocation. This allocation included both the sizing of the reservoir and a determination of the values of the storages to be distributed to the water supply function and the flood control function. We also showed how the allocation of reservoir services could be performed for the case of three functions: water supply, flood control, and hydropower generation.

Although the optimal division of reservoir services for a new reservoir is an interesting and practical problem in some settings, the reality of the current situation in the United States is that few new reservoirs are being built by the major public agencies that operate in this arena. Although new reservoirs are not, in general, being built, the recalibration of the services of existing reservoirs is real, important, and ongoing. Thus, the problem of reallocation is one that deserves attention alongside the problem of allocation.

We focus then in this chapter on reallocation—specifically on the reallocation of services in the context of a reservoir that fills water supply, flood control, and hydropower purposes. We mentioned in Chapter 6 the survey undertaken by the U.S. Army Corps of Engineers (1988) of potential reallocation settings. That survey identified 16 Corps reservoirs that held reallocation potential, and of the 16, 8 reservoirs were dedicated to water supply and flood control functions, and these alone. Thus, a focus on these two functions as a beginning is entirely appropriate.

One reallocation question that is commonly asked is: "Is there a reallocation of flood storage to water storage and water supply (possibly by month) that will not increase expected flood damages and will provide an additional or new water supply?" If such a reallocation exists, the next question might be: "What is the maximum of revenue from water supply less losses that can be provided, given that expected flood damages cannot be allowed to increase?" The "no increase in expected flood damages" means that expected damages after the reallocation cannot be greater than expected damages before the reallocation.

The statement implied by these questions is that the flood damage functions being used in the reallocation process must differ from those that were used at the time of the reservoir's authorization. The functions may differ for several reasons. First, decades may have elapsed and the design storms and thus hydrologic parameters may be different from those used at the time of authorization. Second, the use of the floodplain may have changed. Less economic activity may now be present in the floodplain, or levees may be in place that were not originally planned. Thus, the damage equations, which are functions of the volume(s) of the flood pool, may be different than they were at the time of authorization. To answer the question of whether a reallocation exists that does not increase expected flood losses, we utilize the same notation that we introduced in Chapter 6 and used again in Chapter 8.

Once again, we will pursue the problem formulation in steps; that is, we will derive the supply curve in steps. The supply curve, as in earlier situations, can be derived for a constant release or time-varying release over the 12 months and for a flood storage uniform over the year or variable by month or season of the year. In the first situation in which we derive the supply curve, we set a constant release over the 12 months and we set flood storage uniform over the same time frame. We will then progress through models that successively relax the requirements for constant monthly releases and flood pools constant through the year.

RELIABILITY-CONSTRAINED MODELS FOR REALLOCATING WATER SUPPLY STORAGE AND FLOOD STORAGE

Model 9-a: Constant Release and Constant Flood Pool

The reallocation process, as applied to these two purposes, would seek the supply curve that can be achieved without an increase in flood damage. This goal is augmented by the requirement that the water supply purpose can be honored with a specified reliability. In this case, c, the reservoir capacity is a known value, in contradistinction to Chapters 6 and 8, where c is still an unknown. Thus, the problem can be stated as

$$\text{Maximize} \quad z = 12p_w q - f(v)$$

$$\text{s.t.} \quad s_n \geq s_o \tag{9-1}$$

$$s_t = s_{t-1} - q - w_t + i_t \quad t = 1, 2, \ldots, n \tag{9-2}$$

$$s_t \leq c - v \qquad t = 1, 2, \ldots, n \tag{9-3}$$

$$s_t, w_t \geq 0 \qquad t = 1, 2, \ldots, n$$

$$q, v \geq 0$$

The reliability constraints for a one-month-duration drought are

$$P[S_t \geq 0] \geq \alpha \qquad t = 1, 2, \ldots, n$$

In addition, the constraint that limits expected annual flood damages to their previous level must be added. That is,

$$f(v) \leq D \tag{9-4}$$

where D is the previous expected annual flood damages scaled to account for inflation.

By using the mathematics assembled earlier, the reliability constraint for the one-month-duration drought has a deterministic equivalent of

$$q - s_{t-1} \leq i_t^{1-\alpha} \qquad t = 2, 3, \ldots, n \tag{9-5}$$

and the constraints for droughts of longer duration are

$$2q - s_t \leq (i2)_t^{1-\alpha} \qquad t = 2, 3, \ldots, n \tag{9-6}$$

$$3q - s_t \leq (i3)_t^{1-\alpha} \qquad t = 3, 4, \ldots, n \tag{9-7}$$

$$4q - s_t \leq (i4)_t^{1-\alpha} \qquad t = 4, 5, \ldots, n \tag{9-8}$$

where all parameters are as defined in earlier chapters.

From the investigations of Chapter 7, we know that we cannot predict in advance which duration drought will bind since we do not know in advance the value of the water supply capacity under which we will operate. Thus, we need to write and include the chance constraints for all possible reasonable drought durations that we could consider as having a possibility of binding. Of course, some or one drought duration constraint *will* bind and whichever one(s) does, it will have been included and it can be identified.

The reader notes that this model for reallocation leaves out reservoir costs; these are sunk costs and are no longer optimizable. Also flood damages are now constrained where in the earlier model they were optimized. We make no claim that constraining expected flood damages to be no more than their prior level is a perfect choice. Instead, it is a choice dictated by the history of the project, by the political process and system.

As before (Chapter 6), a supply curve is derived. In the context of an objective of maximum revenue minus costs, the constant monthly amount of water that would be made available is plotted against the price of water. As the price goes up, more water is made available and more storage is dedicated to water supply and less to flood control. The supply curve stops abruptly, as v is made smaller, when $f(v)$, the losses associated with a flood pool of v, reaches D, the expected annual flood damages specified by the initial allocation.

Absent from these models is the reliability of the flood control function, a topic that will be taken up after the reallocation models have been structured.

Model 9-b: Differing Monthly Releases and Constant Flood Pool

In this case, it is appropriate to maximize the revenue from the sale of water less flood losses where the monthly supply quantities are known portions β of the annual water supply.

The model is

$$\text{Maximize} \quad z = p_w Q_A - f(v)$$

$$\text{s.t.} \quad f(v) \leq D \tag{9-9}$$

$$s_n \geq s_o \tag{9-10}$$

$$s_t = s_{t-1} - q_k - w_t + i_t \quad t = 1, 2, \ldots, n; \\ k = t - 12[(t-1)/12] \tag{9-11}$$

$$s_t \leq c - v \quad t = 1, 2, \ldots, n \tag{9-12}$$

$$q_k - \beta_k Q_A = 0 \quad k = 1, 2, \ldots, 12 \tag{9-13}$$

$$s_t, w_t \geq 0 \quad t = 1, 2, \ldots, n$$

$$q_k \geq 0 \quad k = 1, 2, \ldots, 12$$

$$Q_A, v \geq 0$$

plus we add the one-month drought durations constraints:

$$q_k - s_{t-1} \leq i_t^{1-\alpha} \quad t = 2, 3, \ldots, n; \; k = t - 12[(t-1)/12] \tag{9-14}$$

as well as the constraints for droughts of longer duration

$$q_{k+1} + q_{k+2} - s_t \leq (i2)_t^{1-\alpha} \quad t = 2, 3, \ldots, n; \\ k = t - 12[(t-1)/12] \tag{9-15}$$

$$q_{k+1} + q_{k+2} + q_{k+3} - s_t \leq (i3)_t^{1-\alpha} \quad t = 3, 4, \ldots, n; \\ k = t - 12[(t-1)/12] \tag{9-16}$$

$$\sum_{r=k+1}^{k+4} q_r - s_t \le (i4)_t^{1-\alpha} \qquad t = 4, 5, \ldots, n; \\ k = t - 12[(t-1)/12] \qquad (9\text{-}17)$$

where all parameters are as defined earlier.

Once again, the chance constraints associated with all drought durations that might cause binding need to be included in the model because one cannot predict the water supply volume that comes from the optimization process.

Again, we iterate over price to derive a truncated supply curve of annual water that can be made available as a function of the price of water. The truncation is due to the constraint that prevents expected annual flood damages from increasing.

Model 9-c: Constant Monthly Release and Variable Flood Pool

Once again, we derive the supply curve by maximizing revenue less losses by iterating over the price of water, subject to a constraint on flood damages limited to their previous expected value, scaled to the present.

Maximize
$$z = 12p_w q - \sum_{k=1}^{12} f_k(v_k)$$

s.t.
$$\sum_{k=1}^{12} f_k(v_k) \le D \qquad (9\text{-}18)$$

$$s_n \ge s_o \qquad (9\text{-}19)$$

$$s_t = s_{t-1} + i_t - q - w_t \qquad t = 1, 2, \ldots, n \qquad (9\text{-}20)$$

$$s_t \le c - v_k \qquad t = 1, 2, \ldots, n; \qquad (9\text{-}21) \\ k = t - 12[(t-1)/12]$$

$$s_t, w_t \ge 0 \qquad t = 1, 2, \ldots, n$$

$$v_k \ge 0 \qquad k = 1, 2, \ldots, 12$$

$$q \ge 0$$

Again, we add the chance constraints on nonnegative reservoir storages, constraints (9-5) to (9-8).

Note that the water storage volumes, $(c - v_k)$, are variable by month and these values are unknown before the model is run because the v_k are unknown. As a consequence, it is impossible to know in advance which of the duration constraints will bind, and all reasonably plausible duration constraints must be appended so that the binding one will be sure to be included in the analysis.

The last model involves differing monthly releases, keyed to an annual requirement, and a flood pool that varies by month. No new elements, elements we have not considered before, are introduced by this model so it is omitted because its outlines should be clear.

REALLOCATION OF WATER SUPPLY, FLOOD CONTROL, AND HYDROPOWER SERVICES USING THE RELIABILITY-CONSTRAINED MODEL

The reallocation process is extended here to include the hydropower function in addition to water supply and flood control. The results of solving the reallocation formula are water supply release(s), flood pool(s), and firm hydro-energy levels. Capacity, of course, is a given.

In our discussion of hydro, we once again make note of our assumption that water supply releases are separate from releases through the turbine, that the one is not synonymous with the other. Releases through the turbine enter the downstream and become unavailable to M&I water users according to our assumption. As before, we attach a known value to firm energy only and not to dump energy, but dump energy valuation can be structured in the same fashion as in Chapter 6.

The notation in this chapter is the same as that in Chapter 8, so we can proceed directly to the formulation whose solution furnishes the supply curve. Because of the nonlinear nature of the mathematical model of hydropower, direction solution for release(s), flood pool(s), and firm hydroenergy is not possible, but we can, by investigating a range of firm energies d, locate the most profitable division among water supply, flood control, and hydropower production. Our approach will focus on the situation of constant monthly release to water supply and constant flood pool, but obviously these assumptions could be relaxed as in the first main section.

The objective function to be maximized is the difference between the two revenue streams from the sale of water and hydropower and losses associated with floods. The cost of the reservoir no longer enters the analysis unless there are special costs associated with hydro retrofitting.

We seek to

$$\text{Maximize } z = 12p_w q - f(v)$$

subject to constraints (8-23) to (8-30). These are the constraints that define the operation of the reservoir and that require its storage to be nonnegative with a desired level of reliability. Also included is (8-25), which mandates a specified level of hydro-energy delivery, and constraint (9-4), which enforces a continuation of flood-loss prevention at the previous level. The nonlinear hydropower constraint can, for each stated level of firm energy delivery, be linearized. As a consequence, since p_w is specified and $f(v)$ is capable of piecewise approximation, the entire problem can be solved as a linear program—for each specified value of hydroenergy delivery and a specified price of water.

The annual revenue from the sale of that hydroenergy is calculated by multiplying its known price by the firm monthly delivery and summing over the 12 months of the year. This term is added to the objective value above which now represents on an annual basis the net of revenues from both water and hydro less flood losses. The firm delivery is then changed for a reanalysis of the objective function with the price of water still specified as a constant. That is, over a range of firm hydro levels, the most profitable division of reservoir services is chosen—with a single set of firm water supply, flood storage, and firm hydrodelivery as the outputs of the analysis.

This choice represents the most cost-effective way to divide the reservoir's services for a particular price of M&I water supply.

For another price for water, the process of iteration over firm energy levels is repeated and a new optimal division of reservoir services is calculated. This step is repeated over a range of prices of water, each step providing—at that water price—the best division between the services of water supply, hydro, and flood prevention. A supply curve is built up in this way, a curve of water supply available as a function of price, but the curve, as in the first main section, is truncated by the requirement to continue to provide flood prevention at the same level as specified in the past. Calculation of a storage-rule curve, as described in Chapter 6, is appropriate.

FLOOD CONTROL RELIABILITY

This topic, the last to be considered, has been in the background in Chapters 6 and 7 and is finally brought forward for a fuller but still brief consideration. In fact, flood control reliability could use significant study—perhaps along the lines of investigation that were utilized for water supply reliability in Chapter 7. An investigator may wish to pick up this interesting issue for further study. The treatment here, however, is by sensitivity analysis.

Suppose a mathematical reallocation study is successfully completed—in the sense of mathematical computation only—for the case of Model 9-a, constant monthly release, constant flood control volume. The solution to this model provides a new flood volume v, it is hoped smaller than the value in the original authorization, and a new water storage volume $(c - v)$. Given this flood pool volume, call it for the moment v^0, the probability of spillway use and that of dam failure should be evaluated. If either of these are unacceptably large, further analysis is required.

In particular, if either probability is unacceptable, v^0 must be increased, and it would be increased in stages until the probabilities of spillway use and dam failure are acceptably reduced. Of course, expected damages would be decreasing along this path, so the flood damage constraint would not be violated. At this new final volume v^F, water supply capacity is $(c - v^F)$ and the reallocation of reservoir services, subject to the water supply reliability requirements, is calculated. Of course, revenue less losses is now less than it was since the water storage volume has been decreased.

Now suppose a mathematical reallocation study has been completed for the case of Model 9-c, constant monthly release and variable flood pool. A set of flood pool volumes, v_k^0 variable by month k, is one of the outputs from this model. The largest probabilities of spillway use and of dam failure are calculated over the year. If either of these are unacceptably large, further analysis is required. A rough-and-ready way to arrive at the set of 12 v_k that limit the failure probabilities to acceptable levels (and do not violate the expected damage constraint) is to investigate each month in turn in which failure probabilities are too large. In each of these months, increase the values of the v_k in increments just to the levels v_k^F that satisfy allowable failure probability. Calculate the water storage volumes by month; these are ($c - v_k^F$). For these volumes, calculate the reallocation of reservoir services subject to water supply reliability requirements. Of course, the largest feasible release has decreased because of the reduced monthly water supply volumes.

REFERENCES

U.S. Army Corps of Engineers. 1988. *Opportunities for Reservoir Storage Reallocation,* IWR Policy Study 88-PS-2. Water Resources Support Center, Institute for Water Resources, Ft. Belvoir, Va.

INDEX